Javascript for R

Chapman & Hall/CRC
The R Series

Series Editors

John M. Chambers, Department of Statistics, Stanford University, California, USA
Torsten Hothorn, Division of Biostatistics, University of Zurich, Switzerland
Duncan Temple Lang, Department of Statistics, University of California, Davis, USA
Hadley Wickham, RStudio, Boston, Massachusetts, USA

Recently Published Titles

Statistical Inference via Data Science: A ModernDive into R and the Tidyverse
Chester Ismay and Albert Y. Kim

Reproducible Research with R and RStudio, Third Edition
Christopher Gandrud

Interactive Web-Based Data Visualization with R, plotly, and shiny
Carson Sievert

Learn R: As a Language
Pedro J. Aphalo

Using R for Modelling and Quantitative Methods in Fisheries
Malcolm Haddon

R For Political Data Science: A Practical Guide
Francisco Urdinez and Andres Cruz

R Markdown Cookbook
Yihui Xie, Christophe Dervieux, and Emily Riederer

Learning Microeconometrics with R
Christopher P. Adams

R for Conservation and Development Projects: A Primer for Practitioners
Nathan Whitmore

Using R for Bayesian Spatial and Spatio-Temporal Health Modeling
Andrew B. Lawson

Engineering Production-Grade Shiny Apps
Colin Fay, Sébastien Rochette, Vincent Guyader, and Cervan Girard

Javascript for R
John Coene

Advanced R Solutions
Malte Grosser, Henning Bumann, and Hadley Wickham

For more information about this series, please visit: https://www.crcpress.com/
Chapman--HallCRC-The-R-Series/book-series/CRCTHERSER

Javascript for R

John Coene

CRC Press
Taylor & Francis Group
Boca Raton London New York

CRC Press is an imprint of the
Taylor & Francis Group, an **informa** business

A CHAPMAN & HALL BOOK

First edition published 2021
by CRC Press
6000 Broken Sound Parkway NW, Suite 300, Boca Raton, FL 33487-2742

and by CRC Press
2 Park Square, Milton Park, Abingdon, Oxon, OX14 4RN

© 2021 John Coene

CRC Press is an imprint of Taylor & Francis Group, LLC

Library of Congress Cataloging-in-Publication Data

ISBN: 9780367680640 (hbk)
ISBN: 9780367680633 (pbk)
ISBN: 9781003134046 (ebk)

DOI: 10.1201/9781003134046

Typeset in Latin Modern font
by KnowledgeWorks Global Ltd.

Contents

II   Data Visualisation   39

3   Introduction to Widgets   41

4   Basics of Building Widgets   49

5   Your First Widget   57

IV   JavaScript for Computations   253

18 The V8 Engine   255

19 Machine Learning   267

V   Robust JavaScript   273

20 Managing JavaScript   275

VI   Closing Remarks   321

24 Conclusion   323

Bibliography   327

Index   331

List of Figures

Preface

The R programming language has seen the integration of many languages; C, C++, Python, to name a few, can be seamlessly embedded into R so one can conveniently call code written in other languages from the R console. Little known to many, R works just as well with JavaScript–this book delves into the various ways both languages can work together.

The ultimate aim of this work is to demonstrate to readers the many great benefits can reap by inviting JavaScript into their data science workflow. In that respect, the book is not teaching one JavaScript but instead demonstrates how little JavaScript can significantly support and enhance R code. Therefore, the focus is on integrating external JavaScript libraries and only limited knowledge of JavaScript is required in order to learn from the book. Moreover, the book focuses on generalisable learnings so the reader can transfer takeaways from the book to solve real-world problems.

Throughout the book, several Shiny applications and R packages are put together as examples. All of these, along with the code for the entire book, can be found on the GitHub repository: github.com/JohnCoene/javascript-for-r[1].

Premise

The R programming language has been propelled into web browsers with the introduction of packages such as Shiny[2] (Chang et al., 2021a) and rmarkdown[3] (Allaire et al., 2021) which have greatly improved how R users can communicate complex insights by building interactive web applications and interactive documents. Yet most R developers are not familiar with one of web browsers' core technology: JavaScript. This book aims to remedy that by revealing how much JavaScript can greatly enhance various stages of data science pipelines from the analysis to the communication of results.

---

[1] https://github.com/JohnCoene/javascript-for-r

[2] https://shiny.rstudio.com/

[3] https://rmarkdown.rstudio.com/

Notably, the focus of the book truly is the integration of JavaScript with R, where both languages either actively interact with one another, or where JavaScript enables doing things otherwise not accessible to R users. It is not merely about including JavaScript code that works alongside R.

Book Structure

1. The book opens with an introduction to illustrate its premise better it provides rationales for using JavaScript in conjunction with R, which it supports with existing R packages that use JavaScript and are available on CRAN. Then it briefly describes concepts essential to understanding the rest of the book to ensure the reader can follow along. Finally, this part closes by listing the various methods with which one might make JavaScript work with R.

2. We explore existing integrations of JavaScript and R namely by exploring packages to grasp how these tend to work and the interface to JavaScript they provide.

3. A sizeable part of the book concerns data visualisation it plunges into creating interactive outputs with the htmlwidgets package. This opens with a brief overview of how it works and libraries that make great candidates to integrate with the htmlwidgets package. Then a first, admittedly unimpressive, widget is built to look under the hood and observe the inner workings of such outputs to grasp a better understanding of how htmlwidgets work. Next, we tackle a more substantial library that allows drawing arcs between countries on a 3D globe, which we cover in great depth. The last two chapters go into more advanced topics, such as security and resizing.

4. The fourth part of the book details how JavaScript can work with Shiny. Once the basics are out of the way, the second chapter builds the first utility to display notifications programmatically. Then we create a Shiny application that runs an image classification algorithm in the browser. This is then followed by the creation of custom Shiny inputs and outputs. Finally, Shiny and htmlwidgets are (literally) connected by including additional functionalities in interactive visualisations when used with the Shiny framework.

5. Then the book delves into using JavaScript for computations, namely via the V8 engine and Node.js. After a short introduction, chapters

will walk the reader through various examples: a fuzzy search, a time format converter, and some basic natural language operations.

6. Finally, we look at how one can use some of the more modern JavaScript technologies such as Vue, React, and webpack with R—these can make the use of JavaScript more agile and robust.

7. Next the book closes with examples of all the integrations explored previously. This involves recreating (a part of) the plotly package, building an image classifier, adding progress bars to a Shiny application, building an app with HTTP cookies, and running basic machine learning operations in JavaScript.

8. Finally, the book concludes with some noteworthy remarks on where to go next.

Acknowledgement

Many people in the R community have inspired me and provided the knowledge to write this book, amongst them ultimately are Ramnath Vaidyanathan,[4] for his amazing work on the htmlwidgets (Vaidyanathan et al., 2020) package Kent Russell,[5] from whom I have learned a lot via his work on making Vue and React accessible in R and Carson Sievert,[6] for pioneering probably the most popular integration of R and JavaScript with the plotly (Sievert et al., 2021) package.

Early reviewers also shared precious feedback that helped make the book dramatically better, thanks to Maya Gans,[7] Felipe Mattioni Maturana,[8] and Wei Su[9] for thoroughly going through every line of the book.

Thanks to Cody Davis[10] for the cover image of the book.

---

[4] https://github.com/ramnathv/

[5] https://github.com/timelyportfolio

[6] https://github.com/cpsievert

[7] @mayacelium

[8] @felipe_mattioni

[9] @Wei_Su

[10] https://www.davisuko.com/

Part I

Basics and Roadmap

1

Overview

This book starts with a rationale for integrating JavaScript with R and supports it with examples, namely packages that use JavaScript and are available on CRAN. Then, we list the various ways in which one might go about making both languages work together. In the next chapter, we go over prerequisites and a review of concepts fundamental to fully understand the more advanced topics residing in the forthcoming chapters.

1.1 Rationale

Why blend two languages seemingly so far removed from each other? Well, precisely because they are fundamentally different languages that each have their strengths and weaknesses; combining the two allows making the most of their consolidated advantages and circumvents their respective limitations to produce software altogether better for it.

Nevertheless, a fair reason to use JavaScript might be that the thing one wants to achieve in R has already been realised in JavaScript. Why reinvent the wheel when the solution already exists and that it can be made accessible from R? The R package rmapshaper[1] (Teucher and Russell, 2020) by Andy Teucher that integrates mapshaper,[2] a library to edit geo-spatial-related files such as GeoJSON, or TopoJSON. JavaScript is by no means required to make those computations; they could be rewritten solely in R, but that would be vastly more laborious than wrapping the JavaScript API in R as done by the package rmapshaper.

```r
library(rmapshaper)

# get data
```

---

[1] https://github.com/ateucher/rmapshaper
[2] https://github.com/mbloch/mapshaper/

DOI: 10.1201/9781003134046-1

```
data(states, package = "geojsonio")

states_json <- geojsonio::geojson_json(
  states,
  geometry = "polygon",
  group = "group"
)
#> Registered S3 method overwritten by 'geojsonsf':
#>   method          from
#>   print.geojson geojson
#> Assuming 'long' and 'lat' are longitude and latitude, respectively

states_sp <- geojsonio::geojson_sp(states_json)

# print shape file size
print(object.size(states_sp), units = "Mb")
#> 0.4 Mb

# simplify with rmapshaper
states_sm <- rmapshaper::ms_simplify(states_sp, keep = 0.05)

# print reduced size
print(object.size(states_sm), units = "Mb")
#> 0.2 Mb
```

Another great reason is that JavaScript can do things that R cannot, e.g., run in the browser. Therefore, one cannot natively create interactive visualisations with R. Plotly[3] (Sievert et al., 2021) by Carson Sievert packages the plotly JavaScript library[4] to let one create interactive visualisations solely from R code as shown in Figure 1.1.

```
library(plotly)

plot_ly(diamonds, x = ~cut, color = ~clarity, width = "100%")
```

---

[3]https://plotly-r.com/
[4]https://plot.ly/

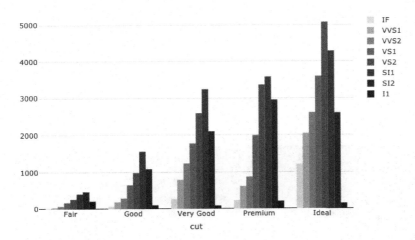

FIGURE 1.1: Basic htmlwidget example

Finally, JavaScript can work together with R to improve how we communicate insights. One of the many ways in which Shiny stands out is that it lets one create web applications solely from R code with no knowledge of HTML, CSS, or JavaScript, but that does not mean they can't extend Shiny–quite the contrary. The waiter package[5] (Coene, 2021b) integrates a variety of JavaScript libraries to display loading screens in Shiny applications as in Figure 1.2.

```r
library(shiny)
library(waiter)

ui <- fluidPage(
  use_waiter(), # include dependencies
  actionButton("show", "Show loading for 3 seconds")
)

server <- function(input, output, session){
  # create a waiter
  w <- Waiter$new()

  # on button click
  observeEvent(input$show, {
    w$show()
    Sys.sleep(3)
```

---

[5]http://waiter.john-coene.com/

```
    w$hide()
  })
}

shinyApp(ui, server)
```

FIGURE 1.2: Waiter screen

Hopefully this makes a couple of great reasons and alluring examples to entice the reader to persevere with this book.

1.2 Methods

Though perhaps not evident at first, all of the packages used as examples in the previous section interfaced with R very differently. As we'll discover, there are many ways in which one can blend JavaScript with R. Generally the way to go about it is dictated by the nature of what is to be achieved.

Let's list the methods available to us to blend JavaScript with R before covering each of them in-depth in their own respective chapter later in the book.

1.2.1   V8

V8[6] by Jeroen Ooms is an R interface to Google's JavaScript engine. It will let you run JavaScript code directly from R and get the result back; it even comes with an interactive console. This is the way the rmapshaper package used in a previous example internally interfaces with the turf.js library.

```
library(V8)

ctx <- v8()

ctx$eval("2 + 2") # this is evaluated in JavaScript!
#> [1] "4"
```

1.2.2   htmlwidgets

htmlwidgets[7] (Vaidyanathan et al., 2020) specialises in wrapping JavaScript libraries that generate visual outputs. This is what packages such as plotly, DT[8] (Xie et al., 2020), highcharter[9] (Kunst, 2020), and many more use to provide interactive visualisation with R.

It is by far the most popular integration out there: at the time of writing it has been downloaded nearly 10 million times from . It will therefore be covered extensively in later chapters.

1.2.3   Shiny

The Shiny framework allows creating applications accessible from web browsers where JavaScript natively runs; it follows that JavaScript can run *alongside* such applications. Often overlooked though, the two can also work *hand-in-hand* as one can pass data from the R server to the JavaScript front end and vice versa. This is how the previously-mentioned package waiter internally works with R.

---

[6]https://github.com/jeroen/v8

[7]http://www.htmlwidgets.org/

[8]https://rstudio.github.io/DT/

[9]http://jkunst.com/highcharter/

1.3   Methods Amiss

Note that there are also two other prominent ways one can use JavaScript with R that are not covered in this book. The main reason being that they require significant knowledge of specific JavaScript libraries, d3.js and React, and while these are themselves advanced uses of JavaScript, their integration with R via the following listed packages are relatively straightforward.

1.3.1   reactR & vueR

ReactR[10] (Inc et al., 2020) is an R package that emulates very well htmlwidgets but specifically for the React framework[11]. Unlike htmlwidgets, it is not limited to visual outputs and also provides functions to build inputs, e.g., a drop-down menu (like `shiny::selectInput`). The reactable package[12] (Lin, 2020) uses reactR to enable building interactive tables solely from R code as shown in Figure 1.3.

```r
reactable::reactable(iris[1:5, ], showPagination = TRUE)
```

Sepal.Length	Sepal.Width	Petal.Length	Petal.Width	Species
5.1	3.5	1.4	0.2	setosa
4.9	3	1.4	0.2	setosa
4.7	3.2	1.3	0.2	setosa
4.6	3.1	1.5	0.2	setosa
5	3.6	1.4	0.2	setosa

1–5 of 5 rows   Previous   **1**   Next

FIGURE 1.3: reactable package example

There is also the package vueR (You and Russell, 2020), which brings some of Vue to R.

---

[10] https://react-r.github.io/reactR/

[11] https://reactjs.org/

[12] https://glin.github.io/reactable/

1.3.2 r2d3

r2d3[13] (Strayer et al., 2020) by RStudio is an R package designed specifically to work with d3.js[14]. It is similar to htmlwidgets but works rather differently, it allows create visualisations such as Figure 1.4.

```
# https://rstudio.github.io/r2d3/articles/gallery/chord/
r2d3::r2d3(
  data = matrix(round(runif(16, 1, 10000)), ncol = 4, nrow = 4),
  script = "chord.js"
)
```

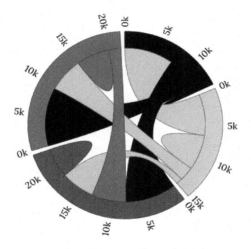

FIGURE 1.4: r2d3 basic example

---

[13]https://rstudio.github.io/r2d3/
[14]https://d3js.org/

2

Prerequisites

The code contained in the following pages is approachable to readers with basic knowledge of R. Still, familiarity with package development using devtools[1] (Wickham et al., 2020c), the Shiny[2] framework (Chang et al., 2021a), the JSON data format, and JavaScript are essential.

The reason for the former is that some of the ways one builds integrations with JavaScript naturally take the form of R packages. Also, R packages make sharing code, datasets, and anything else R-related extremely convenient, they come with a relatively strict structure, the ability to run unit tests, and much more. These have thus become a core feature of the R ecosystem and, therefore, are used extensively in the book as we create several packages. Therefore, the following section runs over the essentials of building a package to ensure everyone can keep up.

Then we briefly go through the JSON data format as it will be used to a great extent to communicate between R and JavaScript. Since both Shiny and JavaScript run in the browser they make for axiomatic companions; we'll therefore use Shiny extensively. Finally, there is an obligatory short introduction to JavaScript.

It is highly recommended to use the freely available RStudio IDE[3] to follow along as it makes a lot of things easier down the line.

2.1 R Package Development

Developing R packages used to be notoriously tricky, but things have considerably changed in recent years, namely thanks to the devtools (Wickham et al., 2020c), roxygen2 (Wickham et al., 2020b), and more recent usethis[4]

---

[1] https://devtools.r-lib.org/
[2] https://shiny.rstudio.com/
[3] https://rstudio.com/products/rstudio/
[4] https://usethis.r-lib.org/

(Wickham and Bryan, 2021) packages. Devtools is short for "developer tools," it is specifically designed to help creating packages; setting up tests, running checks, building and installing packages, etc. The second provides an all too convenient way to generate the documentation of packages, and usethis, more broadly, helps setting up projects, and automating repetitive tasks. Here, we only skim over the fundamentals, there is an entire book by Hadley Wickham called *R Packages*[5] solely dedicated to the topic.

Start by installing those packages from CRAN the roxygen2 package does not need to be explicitly installed as it is a dependency of devtools.

```r
install.packages(c("devtools", "usethis"))
```

2.1.1 Creating a Package

There are multiple ways to create a package. One could manually create every file, use the RStudio IDE, or create it from the R console with the usethis (Wickham and Bryan, 2021) package.

From the RStudio IDE go to File > New Project > New Directory > R Package then select "R package" and fill in the small form, namely name the package and specify the directory where it should be created, as shown in Figure 2.1.

---

[5]http://r-pkgs.had.co.nz/

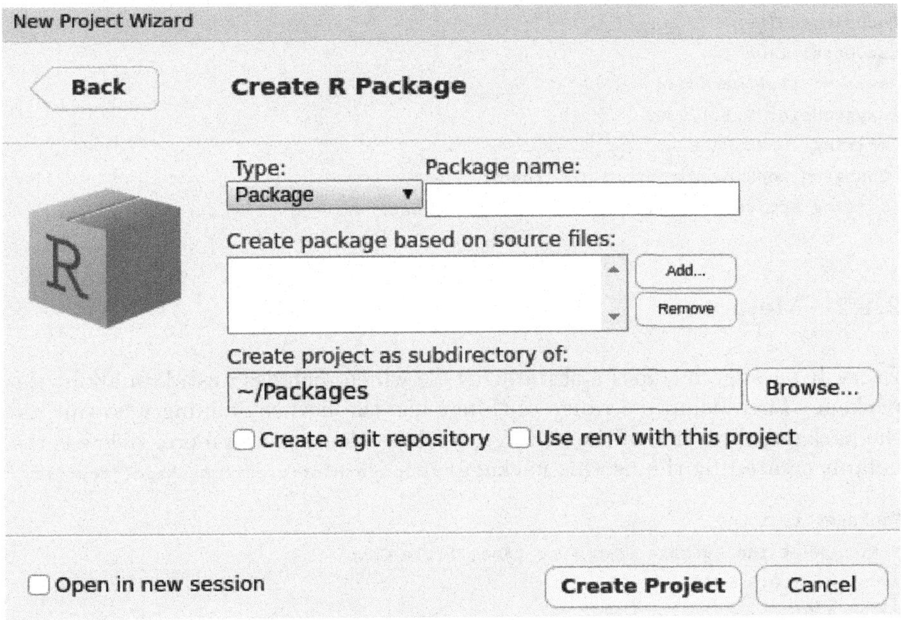

FIGURE 2.1: Package creation wizard

But it could be argued that it's actually more accessible from the R console with the usethis package. The `create_package` function takes as first argument the path to create the package. If you run it from RStudio a new project window should open.

```
# creates a package named "test" in root of directory.
usethis::create_package("test")
```

```
 Creating 'test/'
 Setting active project to '/Packages/test'
 Creating 'R/'
 Writing 'DESCRIPTION'
Package: test
Title: What the Package Does (One Line, Title Case)
Version: 0.0.0.9000
Authors@R (parsed):
    * First Last <first.last@example.com> [aut, cre] (YOUR-ORCID-ID)
Description: What the package does (one paragraph).
License: `use_mit_license()`, `use_gpl3_license()` or friends to
    pick a license
```

```
Encoding: UTF-8
LazyData: true
Roxygen: list(markdown = TRUE)
RoxygenNote: 7.1.1.9000
 Writing 'NAMESPACE'
 Changing working directory to 'test/'
 Setting active project to '<no active project>'
```

2.1.2   Metadata

Every R package includes a DESCRIPTION file which includes metadata about the package. This includes a range of things like the license defining who can use the package, the name of the package, its dependencies, and more. Below is the default created by the usethis package with usethis::create_package("test").

```
Package: test
Title: What the Package Does (One Line, Title Case)
Version: 0.0.0.9000
Authors@R:
    person(given = "First",
           family = "Last",
           role = c("aut", "cre"),
           email = "first.last@example.com",
           comment = c(ORCID = "YOUR-ORCID-ID"))
Description: What the package does (one paragraph).
License: `use_mit_license()`, `use_gpl3_license()` or friends to
    pick a license
Encoding: UTF-8
LazyData: true
Roxygen: list(markdown = TRUE)
RoxygenNote: 7.1.1.9000
```

Much of this is outside the scope of the book. However, it is good to grasp how dependencies are specified. As packages are generally intended for sharing with others, it is vital to ensure users of the package meet the dependencies; otherwise, the package may not work in places. For instance, were we to create a package that relies on one or more functions from the stringr (Wickham, 2019) package we would need to ensure people who install the package have it installed on their machine or those functions will not work.

```
# R/string.R
string_length <- function(string) {
```

```
  stringr::str_length(string)
}
```

Note that the function is preceded by its namespace with :: (more on this later).

The DESCRIPTION file does this; it will make sure that the dependencies of the package are met by users who install it. We can specify such dependencies under Imports, where we can list packages required separated by a comma.

```
Imports:
  stringr,
  dplyr
```

Then again, the usethis package also allows doing so consistently from the R console, which is great to avoid mishandling the DESCRIPTION file.

```
# add stringr under Imports
usethis::use_package('stringr')
```

One can also specify another type of dependencies under Suggests, other packages that enhance the package but are not required to run it. These, unlike package under Imports, are not automatically installed if missing, which can greatly reduce overhead.

2.1.3   R code

An R package must follow a strict structure. R code must be placed in an R/ directory so one should only find .R files in that directory. These files generally contain functions, methods, and R objects.

```
# R/add.R
string_length <- function(strings) {
  stringr::str_length(strings)
}
```

2.1.4 Documentation

Documenting packages used to be notoriously complicated, but thanks to the package roxygen2 that is no longer the case. The documentation of functions of the package (accessible with ?) and datasets that comprise the package reside in separate files sitting in the man/ directory. These are .Rd files that use a custom syntax resembling LaTex. The roxygen package eases the creation of these files by turning special comments and tags in .R files into said .Rd files.

Special comments are a standard R comment # followed by an apostrophe '. The first sentence of the documentation is the title of the documentation file while the second is the description.

```
#' Strings Length
#'
#' Returns the number of characters in strings.
string_length <- function(strings) {
  stringr::str_length(strings)
}
```

There are a plethora of roxygen2 tags to further document different sections. Below we use two different tags to document the parameters and give an example.

```
#' Strings Length
#'
#' Returns the number of characters in strings.
#'
#' @param strings A vector of character strings.
#'
#' @example string_length(c("hello", "world"))
string_length <- function(strings) {
  stringr::str_length(strings)
}
```

As well as generating documentation, the roxygen2 package also allows populating the NAMESPACE file. This is an extensive and often confusing topic, but for this book, we'll be content with the following: the NAMESPACE includes functions that are *imported* and *exported* by the package.

By default, functions that are present in the R files in the R/ directory are not exported: they are not accessible outside the package. Therefore the

`string_length` function defined previously will not be made available to users of
the package, only other functions within the package will be able to call it. To
export it we can use the `@export` tag. This will place the function as exported
in the NAMESPACE file.

```
#' Strings Length
#'
#' Returns the number of characters in strings.
#'
#' @param strings A vector of character strings.
#'
#' @example string_length(c("hello", "world"))
#'
#' @export
string_length <- function(strings) {
  stringr::str_length(strings)
}
```

There are two ways to use external functions (functions from other R packages),
as done thus far in the `string_length` function by using the namespace (package
name) to call the function: `stringr::str_length`. Or by importing the function
needed using a roxygen2 tag thereby removing the need for using the namespace.

```
#' Strings Length
#'
#' Returns the number of characters in strings.
#'
#' @param strings A vector of character strings.
#'
#' @example string_length(c("hello", "world"))
#'
#' @importFrom stringr str_length
#'
#' @export
string_length <- function(strings) {
  str_length(strings) # namespace removed
}
```

Above we import the function `str_length` from the `stringr` package using the
`importFrom` roxygen2 tag. The first term following the tag is the name of the
package wherefrom to import the functions, and the following terms are the
name of the functions separated by spaces so one can import multiple functions

from the same package with, e.g.: `@importFrom stringr str_length str_to_upper`.
If the package imports many functions from a single package one might also
consider importing the package in its entirety with, e.g.: `@import stringr`.

Finally, one can actually generate the `.Rd` documentation files and populate the
NAMESPACE with either the `devtools::document()` function or `roxygen2::roxygenise()`.

 Remember to run `devtools::document()` after changing roxygen2 tags otherwise
changes are not actually reflected in the NAMESPACE and documentation.

2.1.5 Installed files

Here we tackle the topic of installed files as it will be relevant to much of
what the book covers. Installed files are files that are downloaded and copied
as-is when users install the package. This directory will therefore come in very
handy to store JavaScript files that package will require. These files can be
accessed with the `system.file` function, which will look for a file from the root
of the `inst/` directory.

```r
# return path to `inst/dependency.js` in `myPackage`
path <- system.file("dependency.js", package = "myPackage")
```

2.1.6 Build, load, and install

Finally, after generating the documentation of the package with
`devtools::document()` one can install it locally with `devtools::install()`.
This, however, can take a few seconds too many whilst developing a package
as one iterates and regularly tries things; `devtools::load_all()` will not install
the package but load all the functions and object in the global environment to
let you run them.

There is some cyclical nature to developing packages:

1. Write some code
2. Run `devtools::document()` (if documentation tags have changed)
3. Run `devtools::load_all()`
4. Repeat

Note whilst this short guide will help you develop packages good enough for
your system it will certainly not pass checks.

2.2 JSON

JSON (JavaScript Object Notation) is a prevalent data *interchange* format with which we will work extensively throughout this book; it is thus crucial that we have a good understanding of it before we plunge into the nitty-gritty. As one might foresee, if we want two languages to work together, we must have a data format that can be understood by both–JSON lets us harmoniously pass data from one to the other. While it is natively supported in JavaScript, it can be graciously handled in R with the jsonlite package[6] (Ooms, 2020) it is the serialiser used internally by all R packages that we explore in this book.

"To serialise" is just jargon for converting data to JSON.

2.2.1 Serialising

JSON is to all intents and purposes the equivalent of lists in R; a flexible data format that can store pretty much anything–except data.frames, a structure that does not exist in JavaScript. Below we create a nested list and convert it to JSON with the help of jsonlite. We set pretty to TRUE to add indentation for cleaner printing, but this is an argument you should omit when writing production code; it will reduce the file size (fewer spaces = smaller file size).

```r
# install.packages("jsonlite")
library(jsonlite)

lst <- list(
  a = 1,
  b = list(
    c = c("A", "B")
  ),
  d = 1:5
)

toJSON(lst, pretty = TRUE)
#> {
#>   "a": [1],
```

---
[6]https://CRAN.R-project.org/package=jsonlite

```
#>    "b": {
#>       "c": ["A", "B"]
#>    },
#>    "d": [1, 2, 3, 4, 5]
#> }
```

Looking closely at the list and JSON output above, one quickly sees the resemblance. Something seems odd though: the first value in the list (a = 1) was serialised to an array (vector) of length one ("a": [1]), where one would probably expect an integer instead, 1 not [1]. This is not a mistake; we often forget that there are no scalar types in R and that a is, in fact, a vector as we can observe below.

```
x <- 1
length(x)
#> [1] 1
is.vector(x)
#> [1] TRUE
```

JavaScript, on the other hand, does have scalar types; more often than not we will want to convert the vectors of length one to scalar types rather than arrays of length one. To do so we need to use the auto_unbox argument in jsonlite::toJSON; we'll do this most of the time we have to convert data to JSON.

```
toJSON(lst, pretty = TRUE, auto_unbox = TRUE)
#> {
#>    "a": 1,
#>    "b": {
#>       "c": ["A", "B"]
#>    },
#>    "d": [1, 2, 3, 4, 5]
#> }
```

As demonstrated above the vector of length one was "unboxed" into an integer; with auto_unbox set to TRUE, jsonlite will properly convert such vectors into their appropriate type integer, numeric, boolean, etc. Note that this only applies to vectors lists of length one will be serialised to arrays of length one even with auto_unbox turned on: list("hello") will always be converted to ["hello"].

2.2.2 Tabular Data

If JSON is more or less the equivalent of lists in R one might wonder how jsonlite handles dataframes since they do not exist in JavaScript.

```
# subset of built-in dataset
df <- cars[1:2, ]

toJSON(df, pretty = TRUE)
#> [
#>   {
#>     "speed": 4,
#>     "dist": 2
#>   },
#>   {
#>     "speed": 4,
#>     "dist": 10
#>   }
#> ]
```

What jsonlite does internally is essentially turn the data.frame into a list *row-wise* to produce a sub-list for every row then it serialises to JSON. This is generally how rectangular data is represented in lists. For instance, `purrr::transpose` does the same. Another great example is to use `console.table` in the JavaScript console (more on that later) to display JSON data as a table (see Figure 2.2).

```
> console.table([{"speed":4,"dist":2},{"speed":4,"dist":10},{"speed":7,"dist":4},{"speed":7,"dist":22},{"speed":8,"dist":16}])
```

VM1260:1

(index)	speed	dist
0	4	2
1	4	10
2	7	4
3	7	22
4	8	16

▸ Array(5)

FIGURE 2.2: console.table output

We can reproduce this with the snippet below; we remove row names and use apply to turn every row into a list.

```
row.names(df) <- NULL
df_list <- apply(df, 1, as.list)

toJSON(df_list, pretty = TRUE, auto_unbox = TRUE)
#> [
```

```
#>    {
#>      "speed": 4,
#>      "dist": 2
#>    },
#>    {
#>      "speed": 4,
#>      "dist": 10
#>    }
#> ]
```

Jsonlite of course also enables reading data from JSON into R with the function
`fromJSON`.

```
json <- toJSON(df) # convert to JSON
fromJSON(json) # read from JSON
#>   speed dist
#> 1     4    2
#> 2     4   10
```

It's important to note that jsonlite did the conversion back to a data frame.
Therefore the code below also returns a data frame even though the object we
initially converted to JSON is a list.

```
class(df_list)
#> [1] "list"
json <- toJSON(df_list)
fromJSON(json)
#>   speed dist
#> 1     4    2
#> 2     4   10
```

Jsonlite provides many more options and functions that will let you tune how
JSON data is read and written. Also, the jsonlite package does far more than
what we detailed in this section. But at this juncture, this is an adequate
understanding of things.

2.3   JavaScript

The book is not meant to teach JavaScript, only to show how graciously it can work with R. Let us thus go through the very basics to ensure we know enough to get started with the coming chapters.

The easiest way to run JavaScript interactively is probably to create an HTML file (e.g.: `try.html`), write your code within a `<script>` tag and open the file in your web browser. The console output can be observed in the console of the browser, developer tools (see Figure 2.3).

```html
<!-- index.html -->
<html>
  <head>
  </head>
  <body>
    <p id="content">Trying JavaScript!</p>
  </body>
  <script>
    // place your JavaScript code here
    console.log('Hello JavaScript!')
  </script>
</html>
```

FIGURE 2.3: Trying JavaScript

2.3.1   Developer Tools

Most of the JavaScript code written in this book is intended to be run in web browsers; it is thus vital that you have a great understanding of your web browser and its developer tools (devtools). In this section, we discuss those

available in Google Chrome and Chromium, but such tools, albeit somewhat
different, also exist in Mozilla Firefox and Safari.

> The RStudio IDE is built on Chromium, some of these tools will therefore
> also work in RStudio.

The easiest way to access the developer tools from the browser is by "inspecting":
right-click on an element on a webpage and select "inspect." This will open
the developer tools either at the bottom or on the right (Figure 2.4) of the
page depending on the defaults.

FIGURE 2.4: Google Chrome devtools

The developer tools pane consists of several tabs but we will mainly use:

1. Elements: presents the DOM Tree, the HTML document structure,
   great for inspecting the structure of the outputs generated from R.
2. Console: the JavaScript console where messages, errors, and other
   such things are logged. Essential for debugging.

2.3.2   Variable Declaration and Scope

One significant way JavaScript differs from R is that variables must be declared
using one of three keywords, `var`, `let`, or `const`, which mainly affect the scope
where the declared variable will be accessible.

```
x = 1; // error
var x = 1; // works
```

One can declare a variable without assigning a value to it, to then do so later on.

```
var y; // declare
y = [1,2,3]; // define it as array
y = 'string'; // change to character string
```

The let and const keywords were added in ES2015; the const is used to define a constant: a variable that once declared cannot be changed.

```
const x = 1; // declare constant
x = 2; // error
```

Though this is probably only rarely done in R, one can produce something similar by locking the binding for a variable in its .

```
x <- 1 # declare x
lockBinding("x", env = .GlobalEnv) # make constant
x <- 2 # error
unlockBinding("x", env = .GlobalEnv) # unlock binding
x <- 2 # works
```

Notably, const is mainly protecting yourself (the developer) against yourself; if something important is defined and should not change later in the code use const to avoid accidentally reassigning something to it later in the project.

The let keyword is akin to declaring a variable with the var keyword. However, let (and const) will declare the variable in the "block scope." In effect, this further narrows down the scope where the variable will be accessible. A block scope is generally the area within if, switch conditions or for and while loops: areas within curly brackets.

```
if(true){
    let x = 1;
    var y = 1;
```

```
}

console.log(x) // error x does not exist
console.log(y) // works
```

In the above example, x is only accessible within the if statement as it is declared with let, var does not have block scope.

While on the subject of scope, in R like in JavaScript, variables can be accessed from the parent environment (often referred to as "context" in the latter). One immense difference though is that while it is seen as bad practice in R, it is not in JavaScript where it is beneficial.

```
# it works but don't do this in R
x <- 123
foo <- function(){
  print(x)
}
foo()
#> [1] 123
```

The above R code can be re-written in JavaScript. Note the slight variation in the function declaration.

```
// this is perfectly fine
var x = 1;

function foo(){
  console.log(x); // print to console
}

foo();
```

 Accessing variables from the parent (context) is useful in JavaScript but should not be done in R

2.3.3 Document Object Model

One concept does not exist in R is that of the "DOM" which stands for Document Object Model; this is also often referred to as the DOM tree (represented in Figure 2.5) as it very much follows a tree-like structure.

FIGURE 2.5: Document Object Model visualisation

When a web page is loaded, the browser creates a Document Object Model of the web page, which can be accessed in JavaScript from the document object. This lets the developer programmatically manipulate the page itself so one can, for instance, add an element (e.g., a button), change the text of another, and plenty more.

The JavaScript code below grabs the element where id='content' from the document with getElementById and replaces the text (innerText). Even though the page only contains "Trying JavaScript!" when the page is opened (loaded) in the web browser JavaScript runs the code and changes it: this happens very fast so the original text cannot be seen.

```
<!-- index.html -->
<html>
  <head>
  </head>
  <body>
    <p id="content">Trying JavaScript!</p>
  </body>
  <script>
    var cnt = document.getElementById("content");
    cnt.innerText = "The text has changed";
  </script>
</html>
```

One final thing to note for future reference: though not limited to the ids or classes most of such selection of elements from the DOM are done with those where the pound sign refers to an element's id (`#id`) and a dot relates to an element's class (`.class`), just like in CSS.

```html
<!-- index.html -->
<html>
  <head>
  </head>
  <body>
    <p id="content" class="stuff">Trying JavaScript!</p>
  </body>
  <script>
    // select by id
    var x = document.getElementById("content");
    var y = document.querySelector("#content");

    console.log(x == y); // true

    // select by class
    var z = document.querySelector(".stuff");
  </script>
</html>
```

Getting elements from the DOM is a very common operation in JavaScript. A class can be applied to multiple elements, which is useful to select and apply actions to multiple elements. The id attribute must be unique (two elements cannot bear the same id in the HTML document), which is useful to retrieve a specific element.

Interestingly some of that mechanism is used by Shiny to retrieve and manipulate inputs; the argument `inputId` of Shiny inputs effectively defines the HTML id attribute of said input. Shiny can then internally use functions the likes of `getElementById` in order to get those inputs, set or update their values, etc.

```r
shiny::actionButton(inputId = "theId", label = "the label")
```

```html
<button
  id="theId"
  type="button"
  class="btn btn-default action-button">
```

```
    the label
</button>
```

This, of course, only scratches the surface of JavaScript; thus, this provides ample understanding of the language to keep up with the next chapters. Also, a somewhat interesting fact that will prove useful later in the book: the RStudio IDE is actually a browser, therefore, in the IDE, one can right-click and "inspect element" to view the rendered source code.

2.4 Shiny

It is assumed that the reader has basic knowledge of the Shiny framework and already used it to build applications. However, there are some more obscure functionalities that one may not know, but that becomes essential when introducing JavaScript to applications. Chiefly, how to import external dependencies; JavaScript or otherwise.

There are two ways to import dependencies: using the htmltools (Cheng et al., 2021) package to create a dependency object that Shiny can understand, or manually serving and importing the files with Shiny.

2.4.1 Serving Static Files

Static files are files that are downloaded by the clients, in this case, web browsers accessing Shiny applications, as-is. These generally include images, CSS (.css), and JavaScript (.js).

If you are familiar with R packages, static files are to Shiny applications what the inst directory is to an R package; those files are installed as-is. They do not require further processing as opposed to the src folder, which contains files that need compiling, for instance.

There are numerous functions to launch a Shiny application locally; the two most used are probably shinyApp and runApp. The RStudio IDE comes with a convenient "Run" button when writing a shiny application, which when clicked in fact uses the function shiny::runApp in the background. This function looks for said static files in the www directory and makes them available at the same path (/www). If you are building your applications outside of RStudio, you should either also use shiny::runApp or specify the directory which then allows using shiny::shinyApp. Note that this only applies locally; Shiny server

(community and pro) as well as shinyapps.io[7] use the same defaults as the
RStudio IDE and `shiny::runApp`.

To ensure the code in this book can run regardless of the reader's machine or
editor, the asset directory is always specified explicitly (when used). This is
probably advised to steer clear of the potential headaches as, unlike the default,
it'll work regardless of the environment. If you are using golem (Guyader et al.,
2020) to develop your application, then you should not worry about this as it
specifies the directory internally.

Below we build a basic Shiny application. However, before we define the `ui`
and `server`, we use the `shiny::addResourcePath` function to specify the location of
the directory of static files that will be served by the server and thus accessible
by the client. This function takes two arguments: first the `prefix`, which is
the path (URL) at which the assets will be available, second the path to the
directory of static assets.

We thus create the "assets" directory and a JavaScript file called `script.js`
within it.

```r
# run from root of app (where app.R is located)
dir.create("assets")
writeLines("console.log('Hello JS!');", con = "assets/script.js")
```

We can now use the `shiny::addResourcePath` to point to this directory. Generally,
the same name for the directory of static assets and prefix is used to avoid
confusion; below we name them differently for the reader to clearly distinguish
which is which.

```r
# app.R
library(shiny)

# serve the files
addResourcePath(
  # will be accessible at /files
  prefix = "files",
  # path to the assets directory
  directoryPath = "assets"
)

ui <- fluidPage(
  h1("R and JavaScript")
```

---
[7]https://www.shinyapps.io/

```
)

server <- function(input, output){}

shinyApp(ui, server)
```

If you then run the application and open it at the /files/script.js path (e.g.: 127.0.0.1:3000/files/script.js) you should see the content of the JavaScript file (console.log('Hello JS!')), commenting the addResourcePath line will have a "Not Found" error displayed on the page instead.

 All files in your asset directory will be served online and accessible to anyone: do not place sensitive files in it.

Though one may create multiple such directories and correspondingly use addResourcePath to specify multiple paths and prefixes, one will routinely specify a single one, named "assets" or "static," which contains multiple subdirectories, one for each type of static file to obtain a directory that looks something like the tree below. This is, however, an unwritten convention which is by no means forced upon the developer: do as you wish.

```
assets/
├── js/
│   └── script.js
├── css/
│   └── style.css
└── img/
    └── pic.png
```

At this stage, we have made the JavaScript file we created accessible by the clients, but we still have to source this file in the ui as currently this file is, though served, not used by the application. Were one creating a static HTML page one would use the script to src the file in the head of the page.

```
<html>
  <head>
    <!-- source the JavaScript file -->
    <script src="path/to/script.js"></script>
  </head>
  <body>
```

```
  <p id="content">Trying JavaScript!</p>
 </body>
</html>
```

In Shiny we write the UI in R and not in HTML (though this is also supported). Given the resemblance between the names of HTML tags and Shiny UI functions, it is pretty straightforward; the html page above would look something like the Shiny `ui` below.

```
library(shiny)

ui <- fluidPage(
  singleton(
    tags$head(
      tags$script(src = "path/to/script.js")
    )
  ),
  p(id = "content", "Trying JavaScript!")
)
```

The dependency is used in the `htmltools::singleton` function ensures that its content is *only imported in the document once.*

Note that we use the `tags` object, which comes from the Shiny package and includes HTML tags that are not exported as standalone functions. For instance, you can create a `<div>` in Shiny with the `div` function, but `tags$div` will also work. This can now be applied to the Shiny application; the `path/to/script.js` should be changed to `files/script.js`, where `files` is the prefix we defined in `addResourcePath`.

```
# app.R
library(shiny)

# serve the files
addResourcePath(prefix = "files", directoryPath = "assets")

ui <- fluidPage(
  tags$head(
    tags$script(src = "files/script.js")
  ),
  h1("R and JavaScript")
```

```
)

server <- function(input, output){}

shinyApp(ui, server)
```

From the browser, inspecting page (right click > inspect > console tab) one should see Hello JS! in the console, which means the application correctly ran the code in the JavaScript file.

2.4.2 Htmltools

The htmltools package powers much of the Shiny UI, most of the tags that comprise the UI are indeed imported by Shiny from htmltools. For instance shiny::actionButton is just a light wrapper around htmltools tags.

```
shiny::actionButton
```

```
function (inputId, label, icon = NULL, width = NULL, ...)
{
    value <- restoreInput(id = inputId, default = NULL)
    tags$button(
      id = inputId, style = if (!is.null(width))
      paste0("width: ", validateCssUnit(width), ";"),
      type = "button", class = "btn btn-default action-button",
      `data-val` = value, list(validateIcon(icon), label),

      ...
    )
}
```

As the name indicates, htmltools goes beyond the generation of HTML tags and provides broader tools to work with HTML from R; this includes working the dependencies. These may appear simple at first: after all, were one working with an HTML document in order to import HTML or CSS one could use HTML tags.

```
<!-- index.html -->
<html>
  <head>
    <script src="path/to/script.js"></script>
    <link rel="stylesheet" href="path/to/styles.css">
  </head>
  <body></body>
</html>
```

However, it can quickly get out of hand when working with modules and packages. Imagine having to manage the generation of dependencies, such as the above when multiple functions rely on a dependency, but being a dependency, it should only be imported once? The unified framework htmltools helps immensely in dealing with these sorts of issues.

The htmltools package provides utilities to import dependencies and ensure these are only rendered once, as they should be. The way this works is by creating a dependency object that packages like Shiny and R markdown can understand and translate into HTML dependencies. This is handled with the htmlDependency function, which returns an object of class html_dependency.

```
dependency <- htmltools::htmlDependency(
  name = "myDependency",
  version = "1.0.0",
  src = c(file = "path/to/directory"),
  script = "script.js",
  stylesheet = "styles.css"
)
```

About the above, the src argument points to the directory that contains the dependencies (script and stylesheet); this is done with a named vector where file indicates the path is a local directory and href indicates it is a remote server, generally a CDN. Note that one can also pass multiple script and stylesheet by using vectors, e.g.: c("script.js", "anotherScript.js")

CDN stands for Content Delivery Network, a geographically distributed group of servers that provide fast transfer of dependencies.

```r
# dependency to the latest jQuery
dependency <- htmltools::htmlDependency(
  name = "myDependency",
  version = "1.0.0",
  src = c(
    href = "https://cdn.jsdelivr.net/gh/jquery/jquery/dist/"
  ),
  script = "jquery.min.js"
)
```

Shiny, R markdown, and other packages where htmltools is relevant will then be able to translate an `html_dependency` object into actual HTML dependencies. The above would, for instance, generate the following HTML.

```
<script
  src="https://cdn.jsdelivr.net/gh/jquery/jquery/
    dist/jquery.min.js">
</script>
```

Notably, the `htmltools::htmlDependency` also takes a `package` argument, which makes it such that the `src` path becomes relative to the package directory (the `inst` folder). Hence the snippet below imports a file located at `myPackage/inst/assets/script.js`; the ultimate full path will, of course, depend on where the package is installed on the users' machine.

```r
dependency <- htmltools::htmlDependency(
  name = "myDependency",
  version = "1.0.0",
  src = "assets",
  script = c(file = "script.js"),
  package = "myPackage" # user package
)
```

However, how does one use it in R markdown or Shiny? Well, merely placing it in the Shiny UI or an evaluated R markdown chunk will do the job.

```r
# place it in the shiny UI
ui <- fluidPage(
  htmltools::htmlDependency(
```

```
    name = "myDependency",
    version = "1.0.0",
    src = "assets",
    script = c(file = "script.js"),
    package = "myPackage" # user package
  )
)
```

2.4.3 Serving vs. htmltools

For multiple reasons, the best way to include dependencies is probably the former using htmltools. First, it will work with both Shiny and rmarkdown (Allaire et al., 2021) (whereas the other method previously described only works with Shiny), reducing the cognitive load on the developer (you). Learn to use this method and you will be able to import dependencies for many different output types. Moreover, it comes with neat features that will be explored later in the book, e.g., dynamic dependencies for interactive visualisations or Shiny.

Also, using htmltools dependencies will allow other package developers to assess and access the dependencies you build quickly. The function findDependencies will accept another function from which it can extract the dependencies. The object it returns can then be used elsewhere, making dependencies portable. Below we use this function to extract the dependencies of the fluidPage function from the Shiny package.

```
htmltools::findDependencies(
  shiny::fluidPage()
)
```

```
#> [[1]]
#> List of 10
#>  $ name     : chr "bootstrap"
#>  $ version  : chr "3.4.1"
#>  $ src      :List of 2
#>   ..$ href: chr "shared/bootstrap"
#>   ..$ file: chr "/Library/shiny/www/shared/bootstrap"
#>  $ meta     :List of 1
#>   ..$ viewport: chr "width=device-width, initial-scale=1"
#>  $ script   : chr [1:3] "js/bootstrap.min.js"
  "shim/html5shiv.min.js"
  "shim/respond.min.js"
```

```
#>   $ stylesheet: chr "css/bootstrap.min.css"
#>   $ head       : NULL
#>   $ attachment: NULL
#>   $ package    : NULL
#>   $ all_files : logi TRUE
#>   - attr(*, "class")= chr "html_dependency"
```

Extracting dependencies from other packages will become useful later in the book as we assess compatibility between packages: making sure dependencies do not clash and importing dependencies from other packages.

Using `shiny::addResourcePath` has one advantage: its use is not limited to making CSS and JavaScript files available in Shiny; it can be used to serve other file types such as JSON or images that may also be needed in the application.

Part II

Data Visualisation

3

Introduction to Widgets

This part of the book explores the integration of JavaScript with R using the htmlwidgets package, which focuses on libraries that produce a visual output. It is often used for data visualisation but is not limited to it.

As in future parts of this book, we mainly learn through examples, building multiple widgets of increasing complexity as we progress through the chapters. Before writing the first widget, we explore existing R packages that allow creating interactive data visualisations as this gives a first glimpse at what we ultimately build in this part of the book. Then we explore JavaScript libraries that make great candidates for htmlwidgets and attempt to understand how they work to grasp what is expected from the developer in order to integrate them with R. Finally, we build upon the previous chapter to improve how htmlwidgets work with Shiny.

The htmlwidgets package originates from the rCharts (Vaidyanathan, 2013) package created in 2012 by Ramnath Vaidyanathan. It brought together a plethora of data visualisation JavaScript libraries, datamaps, highcharts, morris.js, and many more. Though no longer maintained rCharts ultimately paved the way towards a framework for interactive visualisations in R: two years later, in 2014, Ramnath and other prominent R users start working on htmlwidgets.

The objective of this chapter is to explore existing widgets available on CRAN, discover how they work, whilst focusing on their prominent features as we learn how to implement those in our very own widget in the coming chapters.

3.1 Plotly package

The plotly (Sievert et al., 2021) R package is probably one of the first and the most popular widget built thus far; it has been downloaded from CRAN 4.9 million times at the time of writing this.

DOI: 10.1201/9781003134046-3

Plotly.js is a substantial library that provides over 40 chart types, including 3D charts, statistical graphs, and maps, all of which are available from the R interface. There is so much depth to plotly that there is an entire book[1] on the topic: *Interactive web-based data visualization with R, plotly, and shiny.*

The very short snippet of code below creates Figure 3.1, an interactive scatter plot.

```
library(plotly)

plot_ly(cars, x = ~speed, y = ~dist) %>%
  add_markers()
```

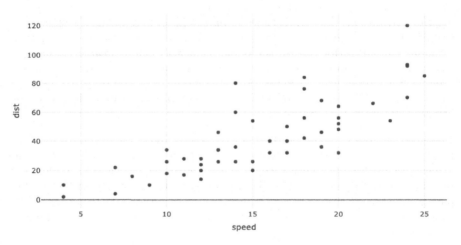

FIGURE 3.1: plotly scatter plot

Plotly also has the ability to translate static ggplot2 (Wickham et al., 2020a) charts to interactive plotly charts with the ggplotly function, as demonstrated in Figure 3.2.

```
p <- ggplot(diamonds, aes(x = log(carat), y = log(price))) +
  geom_hex(bins = 100)
ggplotly(p)
```

---

[1]https://plotly-r.com/

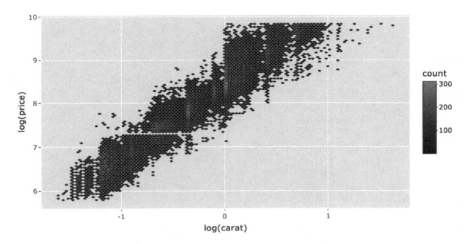

FIGURE 3.2: Interactive ggplot2 with plotly

All plotly charts are initialised with the `plot_ly` function and work nicely with the magrittr (Bache and Wickham, 2020) pipe `%>%`. This implies that (almost) every function expects a plotly object (the output of `plot_ly`) and returns a modified version of that object. The pipe makes code easier to read and more concise.

```
p <- plot_ly(cars, x = ~speed, y = ~dist)
p <- add_markers(p)
```

Plotly implements geoms in a similar fashion as ggplot2, functions that start in `add_` add a layer to the plot (e.g.: `add_lines`, `add_bars`), making it easy to combine series into a single chart, as in Figure 3.3.

```
plot_ly(mtcars, x = ~disp) %>%
  add_markers(y = ~mpg, text = rownames(mtcars)) %>%
  add_lines(y = ~fitted(loess(mpg ~ disp)))
```

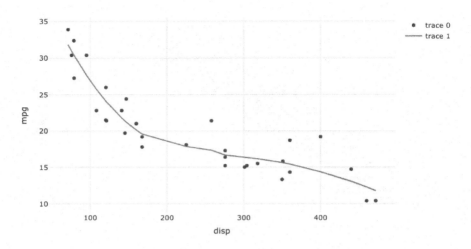

FIGURE 3.3: Multiple layers with plotly

3.2   DT package

The DT package (Xie et al., 2020) is a wrapper for the DataTables[2] jQuery plug-in, which allows creating interactive tables from R `data.frame` objects, it's as simple as a single line of code (see Figure 3.4).

```
DT::datatable(cars)
```

---

[2]https://datatables.net/

	Show 10 ∨ entries		Search: [ ]
		speed ⇕	dist ⇕
1		4	2
2		4	10
3		7	4
4		7	22
5		8	16
6		9	10
7		10	18
8		10	26
9		10	34
10		11	17

Showing 1 to 10 of 50 entries   Previous [1] 2   3   4   5   Next

FIGURE 3.4: Interactive table with DT

DT has grown very popular amongst Shiny developers as it allows capturing server-side many of the users' interactions with the table, such as the selected row, as demonstrated in Figure 3.5.

```r
library(DT)
library(shiny)

ui <- fluidPage(
  fluidRow(
    column(8, DTOutput("myTable")),
    column(
      4,
      h4("Indices of selected rows"),
      verbatimTextOutput("selected")
    )
  )
)

server <- function(input, output) {

  output$myTable <- renderDT({
    datatable(mtcars)
  })

  output$selected <- renderPrint({
    input$myTable_rows_selected
  })
```

```
}

shinyApp(ui, server)
```

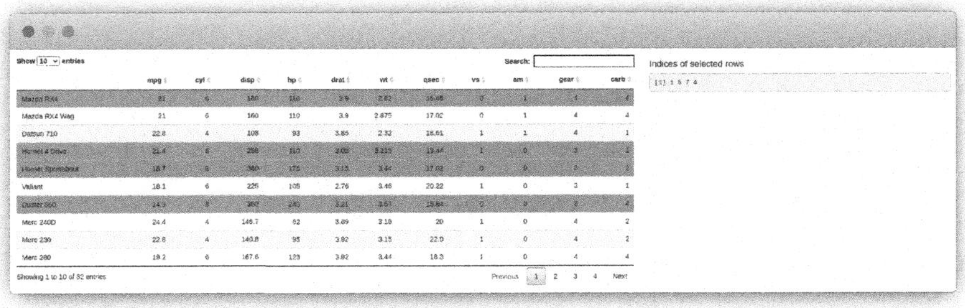

FIGURE 3.5: DT package example

3.3 Crosstalk

DT and plotly both fully support the crosstalk package (Cheng, 2016), a package to enable two widgets to communicate. In the Figure 3.6, we use crosstalk to create a "shared dataset" that can be used to create 1) a table with DT, and 2) a scatter plot with plotly. Rows selected in the table are highlighted in the plot and vice versa.

The `bscols` function is just a convenience function that makes it easy to put the visualisation and table side-by-side.

```
library(DT)
library(plotly)
library(crosstalk)

sd <- SharedData$new(iris[, c("Sepal.Length", "Sepal.Width")])

bscols(
  device = "lg",
  datatable(sd, width = "100%"),
```

```
  plot_ly(sd, x = ~Sepal.Length, y = ~Sepal.Width)
)
```

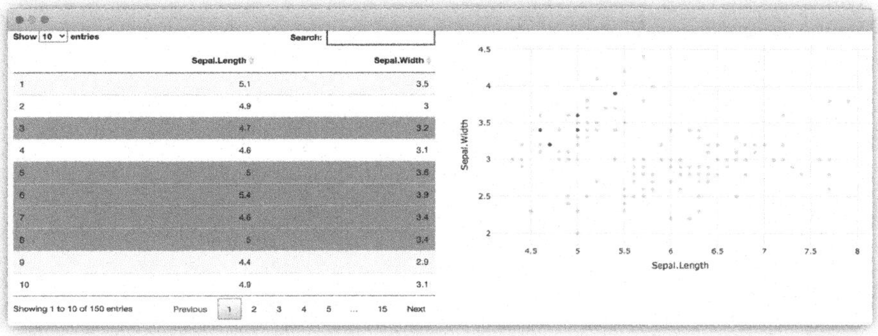

FIGURE 3.6: DT and plotly with crosstalk

3.4 Wrap-up

These are only two examples of widgets and how they can work together. But we hope this makes a compelling case for building such software as we next learn how to build them, integrate them with crosstalk, Shiny, and much more.

4

Basics of Building Widgets

Having explored existing packages that build on top of the htmlwidgets package gives some idea of the end product, but much of how it works and where to start probably remains somewhat mysterious.

4.1 Read and Study

Once you have found an impressive library that you would like to use from R, the very first step when building a widget is to study the JavaScript library you want to integrate thoroughly.

1. Start where the documentation tells you to start, often a "hello world" example, or a get-started section. This will already give you a good sense of how this library functions, the expected data format, and more.
2. Second, it's good to head to the "installation" part of the documentation to know more about the library's dependencies, see whether it is modularised or available in a single bundle, etc.
3. Look at examples in great depth. One cue to know if the library will be more or less complex to integrate with R is whether the various snippets of code that generate the examples are similar or drastically different: commonalities make for easier abstractions and ultimately simpler R code and packages. Some libraries will ultimately be more straightforward to integrate than others.
4. At this stage you should have some idea of what it will take to bring the library to R, and though you likely are vaguely familiar with the functions and methods that comprise the library you must look at the "API," or proper "documentation" of the API to grasp precisely what is available.
5. Finally, before starting the R work, it is advised to properly experience using the library as it was intended: with JavaScript and HTML. This will further give a sense of which parts of the API are

DOI: 10.1201/9781003134046-4

great and which not so much. This is useful to know because you will get to improve or mirror this API in R when you develop the widget for it.

This is likely a section you will want to come back to at the end of this part as you browse for libraries to integrate into your workflow.

4.2   Candidate Libraries

Before going down the rabbit hole, let us explore the types of libraries you will work with; htmlwidgets' main clients so to speak. Below we look at some such popular libraries and briefly analyse how they work and what they have in common. This will significantly help readers conceptualise what trying to achieve in this chapter.

4.2.1   Plotly.js

Plotly.js[1] is probably one of the more popular out there; it provides over 40 fully customisable chart types, many of which are very sophisticated. That is indeed the JavaScript library used by the R package of the same name: plotly.

Looking at the code presented in the "Get Started" guide reveals just how convenient the library is. In Figure 4.1 we import plotly, of course, then have a `<div>` where the visualisation will be placed. Then, using `Plotly.newPlot`, create the actual visualisation by passing it first the element previously mentioned and a JSON of options that describe the chart.

```html
<html xmlns="http://www.w3.org/1999/xhtml" lang="" xml:lang="">
<head>
  <!-- Import library -->
  <script src="plotly-latest.min.js"></script>
</head>
<body>
  <!-- div to hold visualisation -->
  <div id="chart" style="width:600px;height:400px;"></div>
  <!-- Script to create visualisation -->
  <script>
    el = document.getElementById('chart');
```

---

[1]https://plotly.com/javascript/

```
    Plotly.newPlot(el, [{
      x: [1, 2, 3, 4, 5],
      y: [1, 2, 4, 8, 16] }]
    );
  </script>
</body>
</html>
```

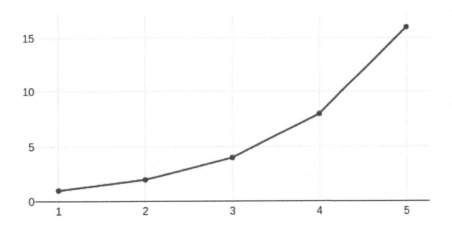

FIGURE 4.1: Plotly example

Now let's look at how another popular library does it.

4.2.2 Highchart.js

Highcharts[2] is another library that allows creating gorgeous visualisation, maps, and more; it's also very popular, albeit not being entirely free.

```
<html xmlns="http://www.w3.org/1999/xhtml" lang="" xml:lang="">
<head>
  <!-- Import library -->
  <script src="highcharts.js"></script>
</head>
```

---

[2]https://www.highcharts.com/

```
<body>
  <!-- div to hold visualisation -->
  <div id="chart" style="width:100%;height:400px;"></div>
  <!-- Script to create visualisation -->
  <script>
    var myChart = Highcharts.chart('chart', {
        xAxis: {
            categories: ['Apples', 'Bananas', 'Oranges']
        },
        series: [{
            name: 'Jane',
            data: [1, 0, 4]
        }, {
            name: 'John',
            data: [5, 7, 3]
        }]
    });
  </script>
</body>
</html>
```

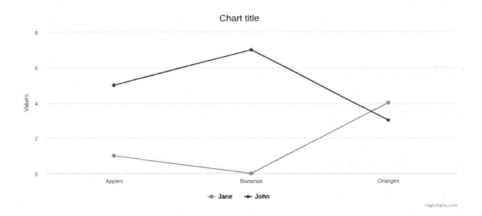

FIGURE 4.2: Highcharts example

Figure 4.2 is very similar to what plotly.js requires: import libraries, create a
<div> where to put the visualisation. Then, to create the chart, run a function
which also takes the id of the div where to place said chart and a JSON of
options defining the actual chart, including the data.

4.2.3 Chart.js

Chart.js[3] is yet another library with which to draw standard charts; it is popular for its permissive license and convenient API.

```html
<html xmlns="http://www.w3.org/1999/xhtml" lang="" xml:lang="">
<head>
  <!-- Import library -->
  <script src="Chart.min.js"></script>
</head>
<body>
  <!-- canvas to hold visualisation -->
  <canvas id="chart"></canvas>
  <!-- Script to create visualisation -->
  <script>
    var el = document.getElementById('chart').getContext('2d');
    var myChart = new Chart(el, {
      type: 'bar',
      data: {
        labels: [
          'Red', 'Blue', 'Yellow', 'Green',
          'Purple', 'Orange'],
        datasets: [{
          label: '# of Votes',
          data: [12, 19, 3, 5, 2, 3]
        }]
      }
    });
  </script>
</body>
</html>
```

---

[3]https://www.chartjs.org/

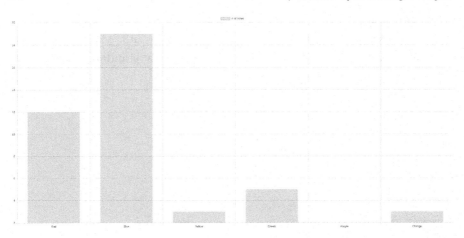

FIGURE 4.3: Chart.js example

In Figure 4.3, we again observe a very similar structure as with previous libraries. The library is imported; instead of a `div` chart.js uses a `canvas`, the visualisation is also created from a single function which takes the canvas as first argument and a JSON of options as second.

Hopefully, this reveals the repeating structure such libraries tend to follow as well as demonstrate how little JavaScript code is involved. It also hints at what should be reproduced, to some extent at least, using R.

4.3 How It Works

Imagine there is no such package as htmlwidgets to help create interactive visualisations from R: how would one attempt to go about it?

As observed, an interactive visualisation using JavaScript will be contained within an HTML document. Therefore it would probably have to be created first. Secondly, the visualisation that is yet to be created likely relies on external libraries; these would need to be imported in the document. The document should also include an HTML element (e.g.: `<div>`) to host said visualisation. Then data would have to be serialised in R and embedded into the document, where it should be read by JavaScript code that uses it to create the visualisation. Finally, all should be managed to work seamlessly across R markdown, Shiny, and other environments.

This gives the basic diagram shown in Figure 4.4; it will be broken down further in the next chapter as the first widget is built.

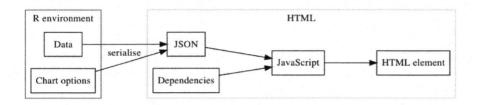

FIGURE 4.4: htmlwidgets inner-workings visualised

Thankfully the htmlwidgets package is there to handle most of this. Nonetheless, it is essential to understand that these operations are undertaken (to some degree) by htmlwidgets.

5

Your First Widget

The previous chapter gave some indication as to how widgets work, but this is overall probably still shrouded in mystery. This chapter aims at demystifying what remains confusing. That is done by building a very basic widget to rummage through its components to observe how they interact and ultimately grasp a greater understanding of how such interactive outputs are actually produced.

5.1 The Scaffold

Though one could probably create widgets outside of an R package, it would only make things more complicated. Htmlwidgets naturally take the form of R packages and are stunningly simple to create. Below we create a package named "playground," which will be used to mess around and explore.

```
usethis::create_package("playground")
```

Then, from the root of the package created, we scaffold a widget which we call "play."

```
htmlwidgets::scaffoldWidget("play")
```

This function puts together the minimalistic structure necessary to implement an htmlwidget and opens play.R, play.js, and play.yaml in the RStudio IDE or the default text editor.

You can scaffold multiple widgets in a single package.

DOI: 10.1201/9781003134046-5   57

These files are named after the widget and will form the core of the package. The R file contains core functions of the R API, namely the `play` function, which creates the widget itself, and the `render*` and `*output` functions that handle the widget in the Shiny server and UI, respectively. The `.js` file contains JavaScript functions that generate the visual output.

```
devtools::document()
devtools::load_all()
```

It might be hard to believe, but at this stage one already has a fully functioning widget ready to use after documenting, and building the package. Indeed, the `play.R` file that was created contains a function named "play," which takes, amongst other arguments, a message.

```
play(message = "This is a widget!")
```

FIGURE 5.1: First widget

Figure 5.1 displays the message in the RStudio "Viewer," or your default browser; indicating the function does indeed create an HTML output. One

can use the button button located in the top right of the RStudio "Viewer" to open the message in the web browser, which can prove very useful to look under the hood of the widgets for debugging.

5.2 The HTML Output

With an out-of-the-box htmlwidgets package, one can start exploring the internals to understand how it works. Let us start by retracing the path taken by the message written in R to its seemingly magical appearance in HTML. The `play` function previously used, takes the `message` wraps it into a list which is then used in `htmlwidgets::createWidget`.

```
# forward options using x
x = list(
  message = message
)
```

Wrapping a string in a list might seem unnecessary, but one will eventually add variables when building a more complex widget, starting with a list makes it easier to add them later on.

To investigate the widget we should look under the hood; the source code of the created (and rendered) output can be accessed in different ways, 1) by right-clicking on the message displayed in the RStudio Viewer and selecting "Inspect element," or 2) by opening the visualisation in your browser using the

button located in the top right of the "Viewer," and in the browser right-click on the message to select "Inspect." The latter is advised as web browsers such as Chrome or Firefox provide much friendlier interfaces for such functionalities. They also come with shortcuts to inspect or view the source code of a page.

Below is a part of the `<body>` of the output of `play("This is a widget!")` obtained with the method described in the previous paragraph.

```
<div id="htmlwidget_container">
  <div
    id="htmlwidget-c21cca0e76e520b46fc7"
    style="width:960px;height:500px;"
    class="play html-widget">
    This is a widget!
  </div>
</div>
```

```
<script
  type="application/json"
  data-for="htmlwidget-c21cca0e76e520b46fc7">
  {"x":{"message":"This is a widget!"},"evals":[],"jsHooks":[]}
</script>
```

One thing the source code of the rendered output reveals is the element (`<div>`) created by the htmlwidgets package to hold the message (the class name is identical to that of the widget, `play`), as well as, below it, in the `<script>` tag, the JSON object which includes the `x` variable used in the `play` function. The `div` created bears a randomly generated `id` which one can define when creating the widget using the `elementId` argument.

```
# specify the id
play("This is another widget", elementId = "myViz")
```

```
<!-- div bears id specified in R -->
<div id="myViz"
  style="width:960px;height:500px;"
  class="play html-widget">
  This is another widget
</div>
```

You will also notice that this affects the `script` tag below it, the `data-for` attribute of which is also set to "myViz;" this indicates that it is used to tie the JSON data to a `div`, essential for htmlwidgets to manage multiple visualisations in R markdown or Shiny for instance. Then again, this happens in the background without the developer (you) having to worry about it.

```
<script type="application/json"
  data-for="myViz">
  {"x":{"message":"This is a widget!"},"evals":[],"jsHooks":[]}
</script>
```

Inspecting the output also shows the dependencies imported, these are placed within the `head` HTML tags at the top of the page.

```
<script src="lib/htmlwidgets-1.5.1/htmlwidgets.js"></script>
<script src="lib/play-binding-0.0.0.9000/play.js"></script>
```

This effectively imports the `htmlwidgets.js` library and the `play.js` file, were the visualisation depending on external libraries they would appear alongside those JavaScript files.

5.3 JavaScript Files

Peaking inside the `play.js` file located at `inst/htmlwidgets/play.js` reveals the following code:

```
// play.js
HTMLWidgets.widget({

  name: 'play',

  type: 'output',

  factory: function(el, width, height) {

    // TODO: define shared variables for this instance

    return {

      renderValue: function(x) {

        // TODO: code to render the widget, e.g.
        el.innerText = x.message;

      },

      resize: function(width, height) {

        // TODO: code to re-render the widget with a new size

      }
```

```
  };
 }
});
```

However convoluted this may appear at first, do not let that intimidate you. The name of the widget (`play`) corresponds to the name used to generate the scaffold, it can also be seen in the `createWidget` function used inside the `play.R` file.

```
htmlwidgets::createWidget(
  name = 'play',
  x,
  width = width,
  height = height,
  package = 'playground',
  elementId = elementId
)
```

This is so htmlwidgets can internally match the output of `createWidget` to its JavaScript function. At this stage, it is probably fair to take a look at the diagram (Figure 5.2) of what is happening.

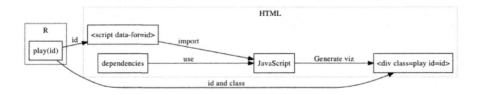

FIGURE 5.2: htmlwidgets internals visualised

The `factory` function returns two functions, `resize` and `renderValue`. The first is used to resize the output dynamically; it is not relevant to this widget and is thus tackled later on. Let us focus on `renderValue`, the function that renders the output. It takes an object `x` from which the "message" variable is used as the text for object `el` (`el.innerText`). The object `x` passed to this function is actually the list of the same name that was built in the R function `play`! While in R one would access the `message` in list `x` with `x$message` in JavaScript to access the `message` in the JSON `x` one writes `x.message`, only changing the dollar sign to a dot. Let us show this perhaps more clearly by printing the content of x.

```
console.log(x);
el.innerText = x.message;
```

We place `console.log` to print the content of x in the console, reload the package with `devtools::load_all` and use the function `play` again, then explore the console from the browser (inspect and go to the "console" tab).

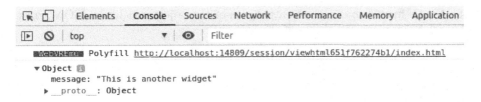

FIGURE 5.3: Console log JavaScript object

Firgure 5.3 displays the JSON object containing the message: it looks eerily similar to the list that was created in R (x = `list(message = "This is a widget!")`). What one should take away from this is that data that needs to be communicated from R to the JavaScript function should be placed in the R list x. This list is serialised to JSON and placed in the HTML output in a `script` tag bearing a `data-for` attribute that indicates which widget the data is destined for. This effectively enables htmlwidgets to match the serialised data with the output elements: data in `<script data-for='viz'>` is to be used to create a visualisation in `<div id='viz'>`.

 Serialisation will make for an important section in a later chapter.

Before we move on to other things one should also grasp a better understanding of the el object, which can also be logged in the console.

```
console.log(x);
console.log(el);
el.innerText = x.message;
```

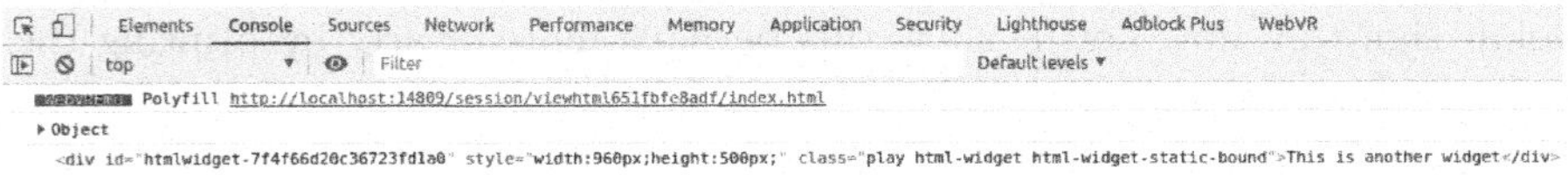

FIGURE 5.4: Console log HTML element

Figure 5.4 displays the HTML element created by htmlwidgets that is meant to hold the visualisation or, in this case, the message. If you are familiar with JavaScript, this is the element that would be returned by `document.getElementById`. This object allows manipulating the element in pretty much any way imaginable: change its position, its colour, its size, or, as done here, insert some text within it. What's more one can access attributes of the object just like a JSON array. Therefore, one can log the `id` of the element.

```
// print the id of the element
console.log(el.id);
el.innerText = x.message;
```

Making the modifications above and reloading the package, one can create a widget given a specific id and see it displayed in the console, e.g.: `play("hello", elementId = "see-you-in-the-console")`.

In an attempt to become more at ease with this setup, let us change something and play with the widget. Out-of-the-box htmlwidgets uses `innerText`, which does very much what it says on the tin, it places text inside an element. JavaScript comes with another function akin to `innerText`, `innerHTML`. While the former only allows inserting text, the latter lets one insert any HTML.

```
el.innerHTML = x.message;
```

After changing the `play.js` file as above, and re-loading the package, one can use arbitrary HTML as messages as shown in Figure 5.5.

```
play("<h1>Using HTML!</h1>")
```

Using HTML!

FIGURE 5.5: First widget output

That makes for a significant improvement, which opens the door to many possibilities. However, the interface this provides is unintuitive. Albeit similar, R users are more familiar with Shiny and htmltools (Cheng et al., 2021) tags than HTML tags, e.g.: `<h1></h1>` translates to `h1()` in R. The package should allow users to use those instead of forcing them to collapse HTML content in a string. Fortunately, there is an effortless way to obtain the HTML from those functions: convert it to a character string.

```r
html <- shiny::h1("HTML tag")

class(html)
#> [1] "shiny.tag"

# returns string
as.character(html)
#> [1] "<h1>HTML tag</h1>"
```

Implementing this in the `play` function will look like this.

```r
# forward options using x
x = list(
  message = as.character(message)
)
```

Reloading the package with `devtools::load_all` lets one use Shiny tags as the message as demonstrated in Figure 5.6.

```r
play(
  shiny::h2("Chocolate is a colour", style = "color:chocolate;")
)
```

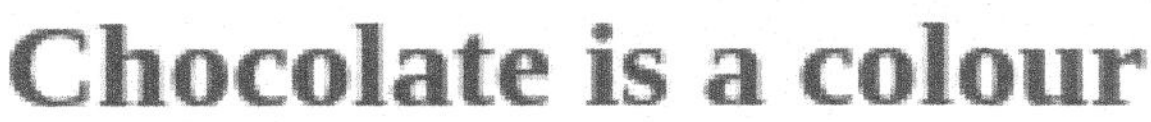

Chocolate is a colour

FIGURE 5.6: First widget using Shiny tags

This hopefully provides some understanding of how htmlwidgets work internally and thereby helps to build such packages. To recapitulate, an HTML document is created in which div is placed and given a specific id; this id is also used in a script tag that contains JSON data passed from R so that a JavaScript function we define can read that data in and use it to generate a visual output in a div. However, as much as this section covered, the topic of JavaScript dependencies was not touched, this is approached in the following section where we build another, more exciting widget, which uses an external dependency.

6

A Realistic Widget

In this section, we build a package called `peity`, which wraps the JavaScript library of the same name, peity.js[1], to create inline charts. This builds upon many things we explored in the playground package built in the previous chapter.

```
usethis::create_package("peity")
htmlwidgets::scaffoldWidget("peity")
```

As done with candidate libraries, as explained in an earlier chapter, there is no avoiding going through the documentation of the library one wants to use to observe how it works. Forging a basic understanding of the library, we can build the following basic example.

```
<!DOCTYPE html>
<html xmlns="http://www.w3.org/1999/xhtml" lang="" xml:lang="">

<head>
  <!-- Import libraries -->
  <script src="jquery-3.5.1.min.js"></script>
  <script src="jquery.peity.min.js"></script>
</head>

<body>
  <!-- div to hold visualisation -->
  <span id="bar">5,3,9,6,5,9,7,3,5,2</span>

  <!-- Script to create visualisation -->
  <script>
    $("#bar").peity("bar");
  </script>
</body>
```

---

[1]https://github.com/benpickles/peity

```
</html>
```

Peity.js depends on jQuery. Hence the latter is imported first; the data for the chart is placed in a `<span>`, and the `peity` method is then used on the element containing the data. Peity.js uses `<span>` HTML tags as these work inline, using a `<div>` the chart will still display, but the purpose of using peity.js would be defeated.

6.1   Dependencies

Once the package is created and the widget scaffold laid down, we need to add the JavaScript dependencies without which nothing can move forward.

Two dependencies are required in order for peity.js to run: peity.js and jQuery. Instead of using the CDN those are downloaded as this ultimately makes the package more robust (more easily reproducible outputs and no requirement for internet connection). Each of the two libraries is placed in its own respective directory.

```
dir.create("./inst/htmlwidgets/jquery")
dir.create("./inst/htmlwidgets/peity")

peity <- paste0(
  "https://raw.githubusercontent.com/benpickles/",
  "peity/master/jquery.peity.min.js"
)
jquery <- paste0(
  "https://code.jquery.com/jquery-3.5.1.min.js"
)

download.file(
  jquery, "./inst/htmlwidgets/jquery/jquery.min.js"
)
download.file(
  peity, "./inst/htmlwidgets/peity/jquery.peity.min.js"
)
```

This produces a directory that looks like this:

```
.
├── DESCRIPTION
├── NAMESPACE
├── R
│   └── peity.R
└── inst
    └── htmlwidgets
        ├── jquery
        │   └── jquery.min.js
        ├── peity.js
        ├── peity.yaml
        └── peity
            └── jquery.peity.min.js
```

In htmlwidgets, dependencies are specified in the `.yaml` file located at `inst/htmlwidgets`, which at first contains a commented template.

```
# (uncomment to add a dependency)
# dependencies:
#  - name:
#    version:
#    src:
#    script:
#    stylesheet:
```

Let's uncomment those lines as instructed at the top of the file and fill it in.

```
dependencies:
  - name: jQuery
    version: 3.5.1
    src: htmlwidgets/jquery
    script: jquery.min.js
  - name: peity
    version: 3.3.0
    src: htmlwidgets/peity
    script: jquery.peity.min.js
```

The order of the dependencies matters. Peity.js depends on jQuery hence the latter comes first in the `.yaml`.

The order in which one specifies the dependencies matters, just like it does in an HTML file, therefore jQuery is listed first. The `stylesheet` entries were removed as none of these libraries require CSS files. The `src` path points to the directory containing the JavaScript files and stylesheets relative to the `inst` directory of the package; this is akin to using the `system.file` function to return the full path to a file or directory within the package.

```
devtools::load_all()
system.file("htmlwidgets/peity", package = "peity")
#> "/home/me/packages/peity/inst/htmlwidgets/peity"
```

We should verify that this is correct by using the one R function the package features and check the source code of the output to verify that the dependencies are well present in the HTML output. We thus run `peity("test")`, open the output in the browser (   ) and look at the source code of the page. At the top of the page, you should see `jquery.min.js` and `jquery.peity.min.js` imported, clicking those links will either present you with the content of the file or an error.

```html
<!DOCTYPE html>
<html>
<head>
<meta charset="utf-8"/>
<style>body{background-color:white;}</style>
<script src="lib/htmlwidgets-1.5.1/htmlwidgets.js"></script>
<script src="lib/jQuery-3.5.1/jquery.min.js"></script>
<script src="lib/peity-3.3.0/jquery.peity.min.js"></script>
<script src="lib/peity-binding-0.0.0.9000/peity.js"></script>
...
```

6.2   Implementation

The JavaScript code for peity.js is relatively uncomplicated. It is just one function, but integrating it with htmlwidgets requires some thinking. In the example below, peity is applied to the element with `id = 'elementId'`; the first argument of `peity` is the type of chart to produce from the data, and the second optional argument is a JSON of options.

```
$("#elementId").peity("bar", {
  fill: ["red", "green", "blue"]
})
```

Also, the data that peity uses to draw the inline chart is not passed to the function but taken from the HTML element.

```
<span id="elementId">5,3,9,6</span>
```

Therefore, the htmlwidget will have to insert the data in the HTML element, then run the `peity` function to render the chart. Inserting the data is actually already done by htmlwidgets by default. Indeed the default htmlwidgets template takes a `message` from the R function, and inserts said message in the HTML element, passing a vector instead of a message produces precisely what peity expects!

```
peity(c(1,5,6,2))
```

```
<div
  id="htmlwidget-495cf47d1a2a4a56c851"
  style="width:960px;height:500px;"
  class="play html-widget">
  1,5,6,2
</div>
```

The argument ought to be renamed nonetheless from `message` to `data`.

```
peity <- function(data, width = NULL, height = NULL,
  elementId = NULL) {

  # forward options using x
  x = list(
    data = data
  )

  # create widget
  htmlwidgets::createWidget(
```

```
    name = 'peity',
    x,
    width = width,
    height = height,
    package = 'peity',
    elementId = elementId
  )
}
```

The change in the R code must be mirrored in the `peity.js` file, where it should set the `innerText` to `x.data` instead of `x.message`.

```
// peity.js
// el.innerText = x.message;
el.innerText = x.data;
```

This leaves the implementation of peity.js to turn the data into an actual chart. The way we shall go about it is to paste one of the examples in the `renderValue` function.

```
renderValue: function(x) {

  // insert data
  el.innerText = x.data;

  // run peity
  $("#elementId").peity("bar", {
    fill: ["red", "green", "blue"]
  })

}
```

One could be tempted to run `devtools::load_all`, but this will not work, namely because the function uses a selector that will not return any object; it needs to be applied to the div created by the widget not `#elementId`. As explained in the previous chapter, the selector of the element created is accessible from the `el` object. As a matter of fact, we did log in the browser console the id of the created div taken from `el.id`. Therefore concatenating the pound sign and the element id produces the selector to said element (`.class`, `#id`).

```
$("#" + el.id).peity("bar", {
  fill: ["red", "green", "blue"]
})
```

This will work but can be further simplified; there is no need to recreate a selector using the `id` of the `el` element; the latter can be used in the jQuery selector directly.

```
$(el).peity("bar", {
  fill: ["red", "green", "blue"]
})
```

This will now produce a working widget, albeit limited to creating charts of a predefined type and colour. Next, these options defining the chart type, fill colours, and others must be made available from R.

Below we add a `type` argument to the `peity` function; this `type` argument is then forwarded to `x`, so it is serialised and accessible in the JavaScript file.

```
peity <- function(data, type = c("bar", "line", "pie", "donut"),
  width = NULL, height = NULL, elementId = NULL) {

  type <- match.arg(type)

  # forward options using x
  x = list(
    data = data,
    type = type
  )

  # create widget
  htmlwidgets::createWidget(
    name = 'peity',
    x,
    width = width,
    height = height,
    package = 'peity',
    elementId = elementId
  )
}
```

This should then be applied by replacing the hard-coded type (`"bar"`) to `x.type`.

```
$(el).peity(x.type, {
  fill: ["red", "green", "blue"]
})
```

Reloading the package will now let one create a chart and define its type, but some options remain hard-coded. These can be made available from R in a variety of ways depending on the interface one wants to provide users of the package. Here we make them available via the three-dot construct (...), which are captured in a list and forwarded to the x object.

```
peity <- function(data, type = c("bar", "line", "pie", "donut"),
  ..., width = NULL, height = NULL, elementId = NULL) {

  type <- match.arg(type)

  # forward options using x
  x = list(
    data = data,
    type = type,
    options = list(...)
  )

  # create widget
  htmlwidgets::createWidget(
    name = 'peity',
    x,
    width = width,
    height = height,
    package = 'peity',
    elementId = elementId
  )
}
```

These can then be easily accessed from JavaScript.

```
$(el).peity(x.type, x.options)
```

This makes (nearly) all of the functionalities of peity.js available from R. Below we use `htmltools::browsable` to create multiple widgets at once, the function only accepts a single value, so the charts are wrapped in an `htmltools::tagList`. Let us explain those in reverse order, `tagList` accepts a group of tags or valid

HTML outputs like htmlwidgets and wraps them into one, it is necessary here because the function `browsable` only accepts one value. Typically htmltools tags are just printed in the console; here we need them to be opened in the browser instead. Remember to run `devtools::load_all` so you can run the `peity` function we just wrote.

```
library(htmltools)

browsable(
  tagList(
    peity(runif(5)),
    peity(runif(5), type = "line"),
    peity("1/4", type = "pie", fill = c("#c6d9fd", "#4d89f9")),
    peity(c(3,5), type = "donut")
  )
)
```

FIGURE 6.1: Peity output with DIV

There is nonetheless one remaining issue in Figure 6.1: peity.js is meant to create inline charts within `<span>` HTML tags but these are created within `<div>` hence each chart appears on a new line.

6.3 HTML Element

As pointed out multiple times, the widget is generated in a `<div>`, which is working fine for most visualisation libraries. However, we saw that peity.js works best when placed in a `<span>` as this allows placing the charts inline.

This can be changed by placing a function named `widgetname_html`, which is looked up by htmlwidgets and used if found. This is probably the first such function one encounters and is relatively uncommon, but it is literally how the htmlwidgets package does it: it scans the namespace of the package looking for a function that starts with the name of the widget and ends in _html and if found uses it. Otherwise, it uses the default div. This function takes the three-dot construct (...) and uses them in an htmltools tag. The three-dots are necessary because internally htmlwidgets need to be able to pass the id, class, and style attributes to the tag.

```
peity_html <- function(...){
  htmltools::tags$span(...)
}
```

This can also come in handy if some arguments must be hard-coded, such as assigning a specific class to every widget.

```
myWidget_html <- function(..., class){
  htmltools::tags$div(..., class = c(class, "my-class"))
}
```

Reloading the package after placing the function above anywhere in the package will produce inline charts, as show in Figure 6.2.

```
browsable(
  tagList(
    p(
      "We can now", peity(runif(5)),
      "use peity", peity(runif(5), type = "line"),
      "inline with text!",
      peity(c(4,2), type = "donut")
    )
  )
)
```

We can now ▮▮▁▮▮ use peity ◺◹ inline with text! ◝

FIGURE 6.2: Peity output with SPAN

7

The Full Monty

With a first widget built, one can jump onto another one: gio.js[1], a library to draw arcs between countries on a three-dimensional globe. This will include many more functionalities such packages can comprise.

Then again, the first order of business when looking to integrate a library is to look at the documentation to understand what should be reproduced in R. Figure 7.1 is a very basic example of using Gio.js in HTML.

```html
<!DOCTYPE html>
<html xmlns="http://www.w3.org/1999/xhtml" lang="" xml:lang="">

<head>
  <!-- Import libraries -->
  <script src="three.min.js"></script>
  <script src="gio.min.js"></script>
</head>

<body>
  <!-- div to hold visualisation -->
  <div id="globe" style="width: 200px; height: 200px"></div>

  <!-- Script to create visualisation -->
  <script>
    var container = document.getElementById("globe");
    var controller = new GIO.Controller(container);
    controller.addData(data);
    controller.init();
  </script>
</body>

</html>
```

---

[1]https://giojs.org/

DOI: 10.1201/9781003134046-7

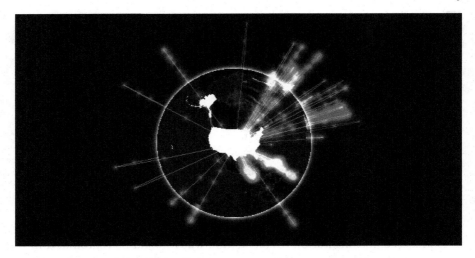

FIGURE 7.1: Gio.js example

Gio.js has itself a dependency, three.js[2], which needs to be imported before gio.js, other than that not much differs from libraries previously explored in this chapter.

```
usethis::create_package("gio")
htmlwidgets::scaffoldWidget("gio")
```

7.1 Dependencies

Handling the dependencies does not differ much, we create the directories within the inst path and download the dependencies within them.

```
# create directories for JS dependencies
dir.create("./inst/htmlwidgets/three", recursive = TRUE)
dir.create("./inst/htmlwidgets/gio", recursive = TRUE)

# download JS dependencies
three <- paste0(
```

---

[2]https://threejs.org/

```
  "https://cdnjs.cloudflare.com/ajax/",
  "libs/three.js/110/three.min.js"
)
gio <- paste0(
  "https://raw.githubusercontent.com/",
  "syt123450/giojs/master/build/gio.min.js"
)

download.file(three, "./inst/htmlwidgets/three/three.min.js")
download.file(gio, "./inst/htmlwidgets/gio/gio.min.js")
```

This should produce the following working directory.

```
.
├── DESCRIPTION
├── NAMESPACE
├── R
│   └── gio.R
└── inst
    └── htmlwidgets
        ├── gio
        │   └── gio.min.js
        ├── gio.js
        ├── gio.yaml
        └── three
            └── three.min.js
```

The libraries have been downloaded, but the gio.yml file is yet to be edited. The order in which the libraries are listed matters. Just as in HTML, three.js needs to precede gio.js as the latter depends on the former and not vice versa.

```
dependencies:
  - name: three
    version: 110
    src: htmlwidgets/three
    script: three.min.js
  - name: gio
    version: 2.0
    src: htmlwidgets/gio
    script: gio.min.js
```

7.2   JavaScript

Let's copy the JavaScript code from the Get Started section of gio.js[3] in the
`gio.js` file's `renderValue` function. At this point, the data format is not known,
so we comment the line, which adds data to the visualisation.

```
// gio.js
HTMLWidgets.widget({

  name: 'gio',

  type: 'output',

  factory: function(el, width, height) {

    // TODO: define shared variables for this instance

    return {

      renderValue: function(x) {

        var container = document.getElementById("globe");
        var controller = new GIO.Controller(container);
        //controller.addData(data);
        controller.init();

      },

      resize: function(width, height) {

        // TODO: code to re-render the widget with a new size

      }

    };
  }
});
```

One can document and load the package, but it will not work as the code

---

[3]https://giojs.org/index.html

above attempts to place the visualisation in a `div` with `id = "globe"`. As for the previously-written widget, this needs to be changed to `el.id`.

```
// gio.js
renderValue: function(x) {

  var container = document.getElementById(el.id);
  var controller = new GIO.Controller(container);
  //controller.addData(data);
  controller.init();

}
```

At this stage, the widget should generate a visualisation (see Figure 7.2).

```
devtools::document()
devtools::load_all()
gio(message = "This required but not used")
```

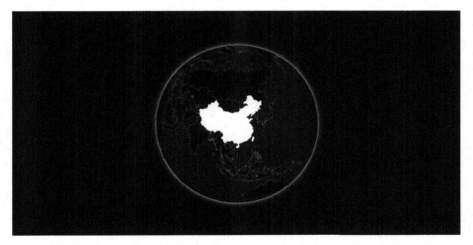

FIGURE 7.2: Gio output without data

Not too shabby given how little work was put into this! Before we move on, let us optimise something. In the JavaScript code, we retrieve the `container` using `el.id`, but this in effect is very inefficient: `el` is identical to `container`.

```
// gio.js
renderValue: function(x) {

  var controller = new GIO.Controller(el);
  //controller.addData(data);
  controller.init();

}
```

7.3 Working with Data

An exciting start, now onto adding data. Let us take a look at the documentation[4] to see what data the library expects.

```
[
  {
    "e": "CN",
    "i": "US",
    "v": 3300000
  },
  {
    "e": "CN",
    "i": "RU",
    "v": 10000
  }
]
```

The JSON data should constitute arrays that denote arcs to draw on the globe where each arc is defined by an exporting country (e), an importing country (i), and is given a value (v). The importing and exporting country, the source and target of the arc, are indicated by ISO alpha-2 country codes. We can read this JSON into R.

```
# data.frame to test
arcs <- jsonlite::fromJSON(
```

---

[4]https://giojs.org/html/docs/dataAdd.html

```
'[
  {
    "e": "CN",
    "i": "US",
    "v": 3300000
  },
  {
    "e": "CN",
    "i": "RU",
    "v": 10000
  }
]'
)
```

```
print(arcs)
#>   e  i      v
#> 1 CN US 3300000
#> 2 CN RU   10000
```

Jsonlite automagically converts the JSON into a data frame where each row is an arc, which is excellent as R users tend to prefer rectangular data over lists: this is probably what the package should use as input too. Let us make some changes to the gio function, so it takes data as input.

```
gio <- function(data, width = NULL, height = NULL, elementId = NULL) {

  # forward options using x
  x = list(
    data = data
  )

  # create widget
  htmlwidgets::createWidget(
    name = 'gio',
    x,
    width = width,
    height = height,
    package = 'gio',
    elementId = elementId
  )
}
```

This must be reflected in the `play.js` file, where we uncomment the line previously commented and use `x.data` passed from R.

```
// gio.js
renderValue: function(x) {

  var controller = new GIO.Controller(el);
  controller.addData(x.data); // uncomment & use x.data
  controller.init();

}
```

We can now use the function with the data to plot arcs!

```
devtools::document()
devtools::load_all()
gio(arcs)
```

Unfortunately, this breaks everything, and we are presented with a blank screen. Using `console.log` or looking at the source code the rendered widget reveals the problem: the data is not actually in the correct format!

```
{
  "x":{
    "data":{
      "e":["CN","CN"],
      "i":["US","RU"],
      "v":[3300000,10000]
    }
  },
  "evals":[],
  "jsHooks":[]
}
```

Htmlwidgets serialised the data frame column-wise (long), where each array is a column, whereas gio.js expect the data to be wide (row-wise serialisation), where each array is an arc (row).

```
# column-wise
jsonlite::toJSON(arcs, dataframe = "columns")
#> {"e":["CN","CN"],"i":["US","RU"],"v":[3300000,10000]}
# row-wise
jsonlite::toJSON(arcs, dataframe = "rows")
#> [{"e":"CN","i":"US","v":3300000},{"e":"CN","i":"RU","v":10000}]
```

As mentioned previously, the convention has it that rectangular data (data frames) are serialised row-wise. That is likely to be a recurring problem for many widgets.

7.4 Transforming Data

There are multiple ways to transform the data and ensure the serialised JSON is as the JavaScript library expects it to be. The following sections explore those various methods before settling on a specific one for the gio package.

7.4.1 Using JavaScript

JavaScript can be used to transform the data, thereby leaving the serialiser as-is only to reshape it into the client. The htmlwidget JavaScript library (already imported by default) exports an object, which provides a method, dataframeToD3, to transform the data from long to wide.

```
// gio.js
renderValue: function(x) {

  // long to wide
  x.data = HTMLWidgets.dataframeToD3(x.data);

  var controller = new GIO.Controller(el);
  controller.addData(x.data);
  controller.init();

}
```

7.4.2   Modify Serialiser

Instead of serialising the data a certain way then correct it in JavaScript as demonstrated previously, one can also modify, or even replace the htmlwidgets default serialiser. Speaking of which, below is the default serializer used by htmlwidgets.

```
function (x, ..., dataframe = "columns", null = "null",
na = "null", auto_unbox = TRUE, use_signif = TRUE,
  digits = getOption("shiny.json.digits", 16), force = TRUE,
  POSIXt = "ISO8601", UTC = TRUE, rownames = FALSE,
  keep_vec_names = TRUE, strict_atomic = TRUE)
{
  if (strict_atomic)
      x <- I(x)
  jsonlite::toJSON(x, dataframe = dataframe, null = null, na = na,
    auto_unbox = auto_unbox, digits = digits, force = force,
    use_signif = use_signif, POSIXt = POSIXt, UTC = UTC,
    rownames = rownames, keep_vec_names = keep_vec_names,
    json_verbatim = TRUE, ...)
}
```

The problem at hand is caused by the `data.frame` argument, which is set to `columns` where it should be set `rows` (for row-wise). Arguments are passed to the serialiser indirectly, in the form of a list set as TOJSON_ARGS attribute to the object x that is serialised. We could thus change the `gio` function to reflect the aforementioned change.

```
gio <- function(data, width = NULL, height = NULL,
  elementId = NULL) {

  # forward options using x
  x = list(
    data = data
  )

  # serialise data.frames to wide (not long as default)
  attr(x, 'TOJSON_ARGS') <- list(dataframe = "rows")

  # create widget
  htmlwidgets::createWidget(
    name = 'gio',
```

```
    x,
    width = width,
    height = height,
    package = 'gio',
    elementId = elementId
  )
}
```

The above may appear confusing at first as one rarely encounters the `attr` replacement function.

```
attr(cars, "hello") <- "world" # set
attr(cars, "hello") # get
#> [1] "world"
```

Other arguments can be placed in the same list; they will ultimately reach the serialiser to modify its output.

7.4.3 Replace Serialiser

Otherwise, the serialiser can also be replaced in its entirety, also by setting an attribute, TOJSON_FUNC, to the x object. Below the serialiser is changed to jsonify (Cooley, 2020), which by default serialises data frames to wide, unlike htmlwidgets' serialiser, thereby also fixing the issue.

```
gio <- function(data, width = NULL, height = NULL,
  elementId = NULL) {

  # forward options using x
  x = list(
    data = data
  )

  # replace serialiser
  attr(x, 'TOJSON_FUNC') <- gio_serialiser

  # create widget
  htmlwidgets::createWidget(
    name = 'gio',
```

```
    x,
    width = width,
    height = height,
    package = 'gio',
    elementId = elementId
  )
}


# serialiser
gio_serialiser <- function(x){
  jsonify::to_json(x, unbox = TRUE)
}
```

7.4.4   Modify the Data

Alternatively, one can also leave all serialisers-related options unchanged and
instead format the data in R prior to the serialisation, changing the dataframe
to a row-wise list.

```
x = list(
  data = apply(data, 1, as.list)
)
```

The above would make it such that the serialiser no longer has to interpret
how the data should be serialised (row-wise or column-wise), the data now
being a list will be serialised correctly.

7.4.5   Pros and Cons

There are pros and cons to each method. The preferable method is probably
to alter the default serialiser **only where needed**; this is the method used in
the remainder of the book. Replacing the serialiser in its entirety should not be
necessary, only do this once you are very familiar with serialisation and truly
see a need for it. Moreover, htmlwidgets' serialiser extends jsonlite to allow
converting JavaScript code, which will come in handy later on. Transforming
the data in JavaScript has one drawback, HTMLWidgets.dataframeToD3 cannot be
applied to the entire x object, it will only work on the subsets that hold the
column-wise data (x.data), which tends to lead to clunky code as one uses said
function in various places.

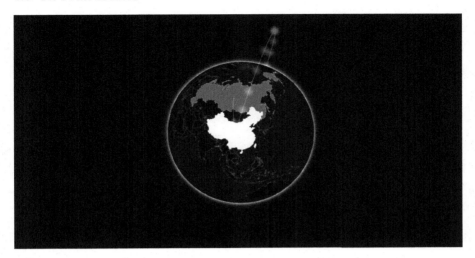

FIGURE 7.3: Gio output with correct serialisation

Figure 7.3 shows that the arcs correctly appear on the globe once the default serialiser has been modified.

7.5 On Print Method

Let us add the option to style the globe: gio.js provides multiple themes[5] but they are currently not applicable from R. As a matter of fact, gio.js provides dozens of customisation options that should be available in the package as well. These, however, probably should be split across different functions, just like they are in gio.js, rather than all be accessible from a single function containing hundreds of arguments. This begs the question, when would one use another function given the function `gio` generates the visualisation? As it happens `gio` itself (or rather the function `htmlwidgets::createWidget` it contains) does not render the output; it returns an object of class "htmlwidget," which renders the visualisation on print (literally `htmlwidget.print` method).

```r
g <- gio(arcs) # nothing renders
g # visualisation renders
```

Therefore, one can use functions on the object returned by `gio`, which contains,

---

[5]https://giojs.org/html/docs/colorStyle.html

amongst other things, the x list to which we can make changes, append data, or do any other operation that standard R lists allow.

```
print(g$x)

## $data
##    e  i      v
## 1 CN US 3300000
## 2 CN RU   10000
##
## attr(,"TOJSON_ARGS")
## attr(,"TOJSON_ARGS")$dataframe
## [1] "rows"
```

This clarified, the function to change the style of the visualisation can be added to the package. It would take as input the output of gio and append the style (name of theme) to the x list; this would then be used in JavaScript to change the look of the visualisation.

```
#' @export
gio_style <- function(g, style = "magic"){
  g$x$style <- style
  return(g)
}
```

The style is applied using the setStyle method on the controller before the visualisation is created, before the init method is called, using the style passed from R: x.style.

```
// gio.js
renderValue: function(x) {

  var controller = new GIO.Controller(el);
  controller.addData(x.data);

  controller.setStyle(x.style); // set style

  controller.init();

}
```

We can now run `devtools::load_all` to export the newly written function and load the functions in the environment with `devtools::load_all` as shown in Figure 7.4.

```
g1 <- gio(arcs)
g2 <- gio_style(g1, "juicyCake")

g2
```

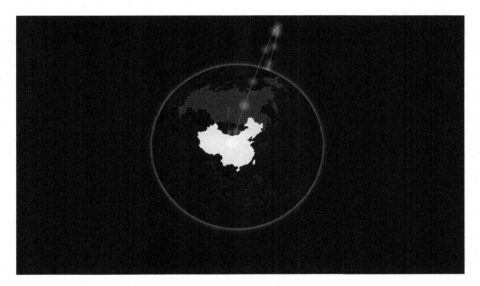

FIGURE 7.4: Gio with a new theme

This is great but can be greatly improved upon with the magrittr pipe (Bache and Wickham, 2020), it would allow effortlessly passing the output of each function to the next to obtain an API akin to that of plotly or highcharter.

```
library(magrittr)

gio(arcs) %>%
  gio_style("juicyCake")
```

The pipe drastically improves the API that gio provides its users and thus probably should be exported by the package; the usethis package provides a function to do so easily.

```
# export the pipe
usethis::use_pipe()
```

This closes this chapter but is not the last we see of gio.js; we shall use it as an example in the next chapters as we extend its functionalities and polish certain aspects such as sizing and security.

8

Advanced Topics

In the previous chapter, we put together an attractive, fully-functioning widget, but it lacks polish and does not use all the features htmlwidgets provides; this chapter explores those. We look into handling the size of widgets to ensure they are responsive as well as discuss potential security concerns and how to address them. Finally, we show how to pass JavaScript code from R to JavaScript and how to add HTML content before and after the widget.

8.1 Shared Variables

Up until now, the topic of shared variables had been omitted as it was not relevant. However, it will be from here onwards. Indeed we are about to discover how to manipulate the widget further; changing the data, resizing, and more. This will generally involve the JavaScript instance of the visualisation, the object named `controller` in the gio package, which, being defined in the `renderValue` function, is not accessible outside of it. To make it accessible outside of `renderValue` requires a tiny but consequential change without which resizing the widget will not be doable, for instance.

The `controller` variable has to be declared outside of the `renderValue` function, inside the `factory`. This was, in fact, indicated from the onset by the following comment: `// TODO: define shared variables for this instance` (generated by `htmlwidgets::scaffoldWidget`). Any variable declared as shown below will be accessible by all functions declared in the `factory`; `renderValue`, but also `resize` and others yet to be added.

```
HTMLWidgets.widget({

  name: 'gio',

  type: 'output',
```

```
factory: function(el, width, height) {

    // TODO: define shared variables for this instance
    var controller;

    return {

        renderValue: function(x) {

            controller = new GIO.Controller(el); // declared outside

            // add data
            controller.addData(x.data);

            // define style
            controller.setStyle(x.style);

            // render
            controller.init();

        },

        resize: function(width, height) {

            // TODO: code to re-render the widget with a new size

        }

    };
  }
});
```

8.1.1 Sizing

The gio function of the package we developed in the previous chapter has
arguments to specify the dimensions of the visualisation (width and height).
However, think how rarely (if ever) one specifies these parameters when
using plotly, highcharter, or leaflet. Indeed HTML visualisations should be
responsive and fit the container they are placed in–not to be confused though;
these are two different things. This enables creating visualisations that look
great on large desktop screens as well as the smaller mobile phones or iPad
screens. Pre-defining the dimensions of the visualisation (e.g.: 400px), breaks

all responsiveness as the width is no longer relative to its container. Using a relative width like 100% ensures the visualisation always fits in the container edge to edge and enables responsiveness.

```r
arcs <- jsonlite::fromJSON(
  '[
    {
      "e": "CN",
      "i": "US",
      "v": 3300000
    },
    {
      "e": "CN",
      "i": "RU",
      "v": 10000
    }
  ]'
)

gio(arcs)
```

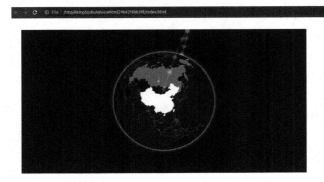

FIGURE 8.1: Gio with no sizing management

When this is not specified, htmlwidgets sets the width of the visualisation to 400 pixels (see Figure 8.1).

```r
gio(arcs, width = 500) # 500 pixels wide
gio(arcs, width = "100%") # fits width
```

These options are destined for the user of the package; the next section details how the developer can define default sizing behaviour.

8.1.2 Sizing Policy

One can specify a sizing policy when creating the widget, the sizing policy will dictate default dimensions and padding in different contexts:

- Global defaults
- RStudio viewer
- Web browser
- R markdown

It is often enough to specify general defaults as widgets are rarely expected to behave differently with respect to size depending on the context, but it can be useful in some cases.

Below we modify the sizing policy of gio via the `sizingPolicy` argument of the `createWidget` function. The function `htmlwidgets::sizingPolicy` has many arguments; we set the default width to 100% to ensure the visualisation fills its container entirely regardless of where it is rendered. We also remove all padding by setting it to 0 and set `browser.fill` to `TRUE`, so it automatically resizes the visualisation to fit the entire browser page.

```r
# create widget
htmlwidgets::createWidget(
  name = 'gio',
  x,
  width = width,
  height = height,
  package = 'gio',
  elementId = elementId,
  sizingPolicy = htmlwidgets::sizingPolicy(
    defaultWidth = "100%",
    padding = 0,
    browser.fill = TRUE
  )
)
```

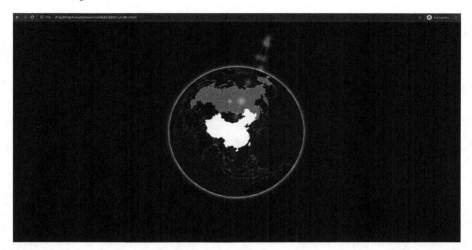

FIGURE 8.2: Gio with sizing policy

Figure 8.2 shows the modified sizingPolicy produces a visualisation that fills the browser.

8.2 Resizing

In the first widget built in this book (playground), we deconstructed the JavaScript factory function but omitted the resize function. The resize function does what it says on the tin: it is called when the widget is resized. What this function will contain entirely depends on the JavaScript library one is working with. Some are very easy to resize, other less so, that is for the developer to discover in the documentation of the library. Some libraries, like gio, do not even require using a resizing (see 8.3) function and handle that automatically under the hood; resize the width of the RStudio viewer or web browser, and gio.js resizes too. This said, there is a function to force gio to resize. Though it is not in the official documentation, it can be found in the source code: resizeUpdate is a method of the controller and does not take any argument.

```
...
resize: function(width, height) {
  controller.resizeUpdate();
```

```
}
...
```

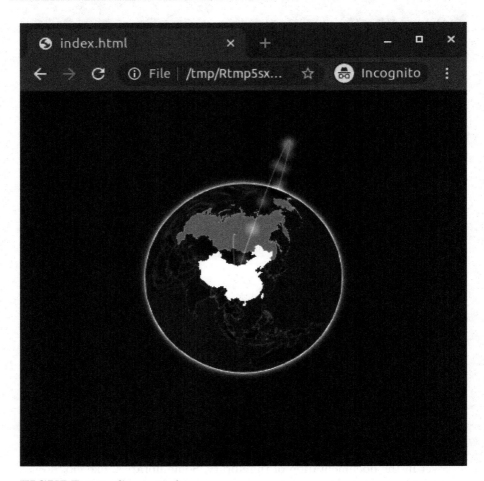

FIGURE 8.3: Gio resized

To give the reader a better idea of what these tend to look like below are the ways plotly, highcharts, and chart.js do it.

Plotly

```
Plotly.relayout('chartid', {width: width, height: height});
```

Highcharts

```
chart.setSize(width, height);
```

Chart.js

```
chart.resize();
```

Note that the `width` and `height` used in the functions above are obtained from the `resize` function itself (see arguments).

That is one of the reasons for ensuring the instance of the visualisation (`controller` in this case) is shared (declared in `factory`). If declared in the `renderValue` function then the `resize` function cannot access that object and thus cannot run the function required to resize the widget.

8.3 Pre Render Hooks and Security

The `createWidget` function also comes with a `preRenderHook` argument, which accepts a function that is run just before the rendering of the widget (in R, not JavaScript), this function should accept the entire widget object as input and should return a modified widget object. That was not used in any of the widgets previously built but is extremely useful. It can be used to make checks on the object to ensure all is correct, or remove variables that should only be used internally, and much more.

Currently, `gio` takes the data frame `data` and serialises it in its entirety which will cause security concerns as all the data used in the widget is visible in the source code of the output. What if the data used for the visualisation contained an additional column with sensitive information? We ought to ensure gio only serialises the data necessary to produce the visualisation.

```
# add a variable that should not be shared
arcs$secret_id <- 1:2
```

We create a `render_gio` function, which accepts the entire widget, filters only the column necessary from the data and returns the widget. This function is then passed to the argument `preRenderHook` of the `htmlwidgets::createWidget` call. This way, only the columns e, v, and i of the data are kept, thus the `secret_id` column will not be exposed publicly.

```r
# preRenderHook function
render_gio <- function(g){
  # only keep relevant variables
  g$x$data <- g$x$data[,c("e", "v", "i")]
  return(g)
}


# create widget
htmlwidgets::createWidget(
  name = 'gio',
  x,
  width = width,
  height = height,
  package = 'gio',
  elementId = elementId,
  sizingPolicy = htmlwidgets::sizingPolicy(
    defaultWidth = "100%",
    padding = 0,
    browser.fill = TRUE
  ),
  preRenderHook = render_gio # pass renderer
)
```

Moreover, security aside, this can also improve performances as only the data relevant to the visualisation is serialised and subsequently loaded by the client. Without the modification above, were one to use gio on a dataset with 100 columns all would have been serialised, thereby significantly impacting performances both of the R process rendering the output and the web browser viewing the visualisation.

8.4   JavaScript Code

As mentioned in a previous chapter, JavaScript code cannot be serialised to JSON.

```r
# serialised as string
jsonlite::toJSON("var x = 3;")
#> ["var x = 3;"]
```

Nonetheless, it is doable with htmlwidgets' serialiser (and only that one). The function `htmlwidgets::JS` can be used to mark a character vector so that it will be treated as JavaScript code when evaluated in the browser.

```
htmlwidgets::JS("var x = 3;")
#> [1] "var x = 3;"
#> attr(,"class")
#> [1] "JS_EVAL"
```

This can be useful where the library requires the use of callback functions, for instance.

Replacing the serialiser will break this feature.

8.5 Prepend and Append Content

There is the ability to append or prepend HTML content to the widget (Shiny, htmltools tags, or a list of those). For instance, we could use `htmlwidgets::prependContent` to allow displaying a title to the visualisation, as shown in Figure 8.4.

```
#' @export
gio_title <- function(g, title){
  title <- htmltools::h3(title)
  htmlwidgets::prependContent(g, title)
}
```

```
gio(arcs) %>%
  gio_title("Gio.js htmlwidget!")
```

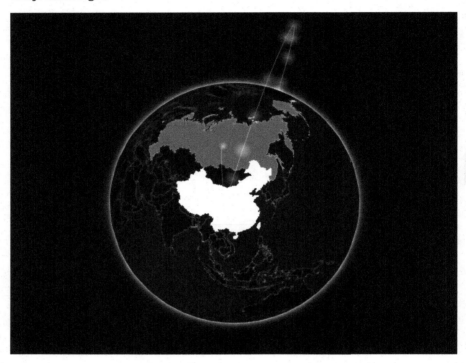

FIGURE 8.4: Gio output with title

While the `prependContent` function places the content above the visualisation, the `appendContent` function places it below, as they accept any valid htmltools or Shiny tag they can also be used for conditional CSS styling for instance.

`prependContent` and `appendContent` do not work in Shiny.

8.6 Dependencies

Thus far, this book has only covered one of two ways dependencies can be included in htmlwidgets. Though the one covered, using the .yml file will likely be necessary for every widget it has one drawback: all dependencies listed in the file are always included with the output. Dependencies can significantly affect the load time of the output (be it a standalone visualisation, an R markdown document, or a Shiny application) as these files may be large. Most large visualisation libraries will therefore allow bundling those dependencies in separate files. For instance, ECharts.js provides a way to customise the bundle to only include dependencies for charts that one wants to draw (e.g., bar chart, or boxplot), highcharts also allows splitting dependencies so one can load those needed for maps, stock charts, and more, separately. It is thus good practice to do the same in widgets, so only the required dependencies are loaded, e.g.: when the user produces a map, only the dependency for that map is loaded. It is used in the leaflet package to load map tiles, for instance.

The Google Chrome network tab (see Figure 8.5) shows the information on resources downloaded by the browser (including dependencies) including how long it takes. It is advisable to take a look at it to ensure no dependency drags load time.

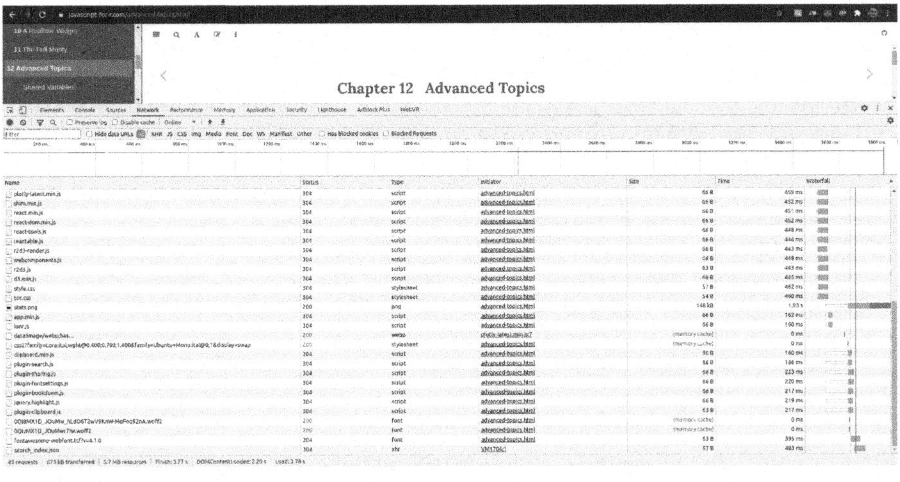

FIGURE 8.5: Google Chrome network tab

To demonstrate, we will add a function in gio to optionally include stats.js[1], a JavaScript performance monitor which displays information such as the number of frames per second (FPS) rendered, or the number of milliseconds needed to render the visualisation. Gio.js natively supports stats.js, but the dependency needs to be imported, and that option needs to be enabled on the `controller` as shown in the documentation[2].

```
// enable stats
controller.enableStats();
```

In htmlwidgets those additional dependencies can be specified via the `dependencies` argument in the `htmlwidgets::createWidget` function or they can be appended to the output of that function.

```
# create gio object
g <- gio::gio(arcs)

is.null(g$dependencies)
```

```
[1] TRUE
```

As shown above, the object created by `gio` includes dependencies, currently NULL as no such extra dependency is specified. One can therefore append those to that object in a fashion similar to what the `gio_style` function does.

From the root of the gio package, we create a new directory for the stats.js dependency and download the latest version from GitHub.

```
dir.create("htmlwidgets/stats")
url <- paste0(
  "https://raw.githubusercontent.com/mrdoob/",
  "stats.js/master/build/stats.min.js"
)
download.file(url, destfile = "htmlwidgets/stats/stats.min.js")
```

First we use the `system.file` function to retrieve *the path to the directory* contains the dependency (`stats.min.js`). It's important that it is the path to the directory and not the file itself.

---

[1] https://github.com/mrdoob/stats.js/
[2] https://giojs.org/html/docs/interfaceStats.html

```
# stats.R
gio_stats <- function(g){

  # create dependency
  path <- system.file("htmlwidgets/stats", package = "gio")

  return(g)

}
```

Then we use the htmltools package to create a dependency, the
htmltools::htmlDependency function returns an object of class html_dependency,
which htmlwidgets can understand and subsequently insert in the output. On
the src parameter, since we reference a dependency from the filesystem we
name the character string file, but we could use the CDN (web-hosted file)
and name it href instead.

```
# stats.R
gio_stats <- function(g){

  # create dependency
  path <- system.file("htmlwidgets/stats", package = "gio")
  dep <- htmltools::htmlDependency(
    name = "stats",
    version = "17",
    src = c(file = path),
    script = "stats.min.js"
  )

  return(g)

}
```

The dependency then needs to be appended to the htmlwidgets object.

```
# stats.R
gio_stats <- function(g){

  # create dependency
  path <- system.file("htmlwidgets/stats", package = "gio")
  dep <- htmltools::htmlDependency(
```

```
    name = "stats",
    version = "17",
    src = c(file = path),
    script = "stats.min.js"
  )

  # append dependency
  g$dependencies <- append(g$dependencies, list(dep))

  return(g)

}
```

Finally, we pass an additional variable in the list of options (x), which we will use JavaScript-side to check whether stats.js must be enabled.

```
#' @export
gio_stats <- function(g){

  # create dependency
  path <- system.file("htmlwidgets/stats", package = "gio")
  dep <- htmltools::htmlDependency(
    name = "stats",
    version = "17",
    src = c(file = path),
    script = "stats.min.js"
  )

  # append dependency to gio.js
  g$dependencies <- append(g$dependencies, list(dep))

  # add stats variable
  g$x$stats <- TRUE

  return(g)
}
```

Then it is a matter of using the stats variable added to x in the JavaScript renderValue function to determine whether the stats feature should be enabled.

```
// gio.js
if(x.stats)
  controller.enableStats();

controller.init();
```

Then the package can be documented to export the newly-created function and loaded in the environment to test the feature, as shown in FIgure 8.6.

```
# create gio object
arcs %>%
  gio() %>%
  gio_stats()
```

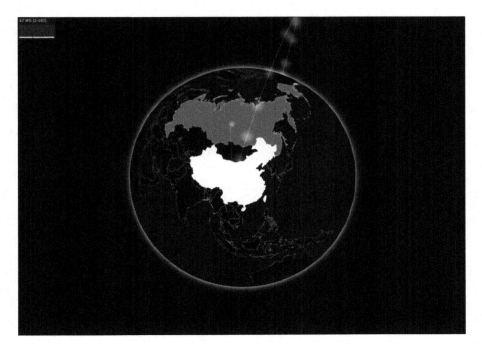

FIGURE 8.6: Gio with stats output

In brief, it is better to only place the hard dependencies in the .yml file; dependencies that are necessary to produce the visualisation and use dynamic dependencies where ever possible. Perhaps one can think of it as the difference between Imports and Suggests in an R package DESCRIPTION file.

8.7 Compatibility

One issue that might arise is that of compatibility between widgets. What if someone else builds another htmlwidget for gio.js uses a different version of the library and that a user decides to use both packages in a Shiny app or R markdown document? Something is likely to fail as two different versions of gio.js are imported, and that one overrides the other. For instance, the package echarts4r (Coene, 2021a) allows working with leaflet but including the dependencies could clash with the leaflet package itself. Therefore, it uses the dependencies from the leaflet package instead.

The htmlwidgets package comes with a function to extract the dependencies from a widget, so they can be reused in another. The function `htmlwidgets::getDependency` returns a list of objects of class `html_dependency`, which can therefore be used in other widgets as demonstrated in the previous section.

```r
# get dependencies of the gio package
htmlwidgets::getDependency("gio")[2:3]
```

```
#> [[1]]
#> List of 10
#>  $ name      : chr "three"
#>  $ version   : chr "110"
#>  $ src       :List of 1
#>   ..$ file: chr "/home/usr/gio/htmlwidgets/three"
#>  $ meta      : NULL
#>  $ script    : chr "three.min.js"
#>  $ stylesheet: NULL
#>  $ head      : NULL
#>  $ attachment: NULL
#>  $ package   : NULL
#>  $ all_files : logi TRUE
#>  - attr(*, "class")= chr "html_dependency"
#>
#> [[2]]
#> List of 10
#>  $ name      : chr "gio"
#>  $ version   : chr "2"
#>  $ src       :List of 1
#>   ..$ file: chr "/home/usr/gio/htmlwidgets/gio"
#>  $ meta      : NULL
#>  $ script    : chr "gio.min.js"
```

```
#>   $ stylesheet: NULL
#>   $ head      : NULL
#>   $ attachment: NULL
#>   $ package   : NULL
#>   $ all_files : logi TRUE
#>   - attr(*, "class")= chr "html_dependency"
```

8.8 Unit Tests

The best way to write unit tests for htmlwidgets is to test the object created by htmlwidgets::createWidget. We provide the following example using testthat (Wickham, 2020), running expect* functions on the output of gio.

```
library(gio)
library(testthat)

test_that("gio has correct data", {
  g <- gio(arcs)

  # internally stored as data.frame
  expect_is(g$x$data, "data.frame")

  # gio does not work without data
  expect_error(gio())
})
```

8.9 Performances

A few hints have already been given to ensure one does not drain the browser; consider assessing the performances of the widget as it is being built. Always try and imagine what happens under the hood of the htmlwidget as you build it; it often reveals potential bottlenecks and solutions.

Remember that data passed to htmlwidgets::createWidget is 1) loaded into R, 2) serialised to JSON, 3) embedded into the HTML output, 4) read back in with JavaScript, which adds some overhead considering it might be read into JavaScript directly. This will not be a problem for most visualisations but might become one when that data is large. Indeed, there are sometimes more efficient ways to load data into web browsers where it is needed for the visualisation.

Consider for instance, geographic features (topoJSON and GeoJSON), why load them into R if it is to then re-serialise it to JSON?

Also, keep the previous remark in mind when repeatedly serialising identical data objects, GeoJSON is again a good example. A map used twice or more should only be serialised once or better not at all. Consider providing other ways for the developer to make potentially large data files accessible to the browser.

Below is an example of a function that could be used within R markdown or Shiny UI to load data in the front end and bypass serialisation. Additionally, the function makes use of AJAX (Asynchronous JavaScript And XML) to asynchronously load the data, thereby further reducing load time.

```r
# this would placed in the shiny UI
load_json_from_ui <- function(path_to_json){
  script <- paste0("
    $.ajax({
        url: '", path_to_json, "',
        dataType: 'json',
        async: true,
        success: function(data){
          console.log(data);
          window.globalData = data;
        }
      });"
    )
  shiny::tags$script(
    script
  )
}
```

Using the above the data loaded would be accessible from the htmlwidgets JavaScript (e.g.: `gio.js`) with `window.globalData`. The `window` object is akin to the `document` object, while the latter pertains to the Document Object Model (DOM) and represents the page, the former pertains to the Browser Object Model (BOM) and represents the browser window. While `var x;` will only be accessible within the script where it is declared, `window.x` will be accessible anywhere.

Note this means the data is read from the web browser, and therefore the data must be accessible to the web browser; the `path_to_json` must thus be a served static file, e.g.: `www` directory in Shiny.

9

Linking Widgets

Widgets can be linked with one another using the crosstalk (Cheng, 2016) package, a fantastic add-on for htmlwidgets that implements inter-widget interactions, namely selection and filtering. This, in effect, allows the selection or filtering of data points in one widget to be mirrored in another. This is enabled with the creation of "shared datasets" that can be used across widgets: outputs that share datasets share interactions.

Crosstalk provides a straightforward interface to the users and instead requires effort from developers for their widgets to support shared datasets.

9.1 Crosstalk Examples

Both the plotly and DT packages support crosstalk, therefore using a shared dataset we can produce a scatter plot with the former and a table with the latter, so that selection of data in one is reflected in the other.

As alluded to earlier on, this can be achieved by using a shared dataset, which can be created with the SharedData R6 class from the crosstalk package. This dataset is then used as one would use a standard dataframe in plotly and DT. The bscols function is just a helper to create columns from HTML elements (see Figure 9.1). It is ideal for examples, but one should not have to use it in Shiny—crosstalk will work without bscols.

```
library(DT)
library(plotly)
library(crosstalk)

shared <- SharedData$new(cars)

bscols(
  plot_ly(shared, x = ~speed, y=~dist),
```

DOI: 10.1201/9781003134046-9

```
  datatable(shared, width = "100%")
)
```

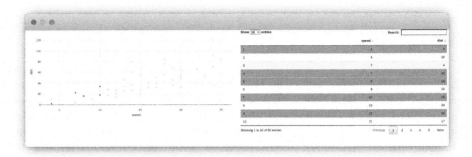

FIGURE 9.1: Crosstalk example

Basic usage of crosstalk datasets in shiny (Figure 9.2) is also straightforward since it accepts reactive expressions to create shared datasets. Note that it takes the expression itself (`expression`) not the output of the expression (`expression()`); the crosstalk documentation explains it best:

> If this feels foreign to you, think of how you pass a function name, not a function call, to `lapply`; that's exactly analogous to what we're doing here.
>
> — Official crosstalk documentation

```
library(DT)
library(shiny)
library(plotly)
library(crosstalk)

ui <- fluidPage(
  selectInput(
    "specie", "Specie",
```

```
    choices = c("setosa", "versicolor", "virginica")
  ),
  fluidRow(
    column(6, DTOutput("table")),
    column(6, plotlyOutput("plot"))
  )
)

server <- function(input, output) {
  reactive_data <- reactive({
    iris[iris$Species == input$specie, ]
  })

  sd <- SharedData$new(reactive_data)

  output$table <- renderDT({
    datatable(sd)
  }, server = FALSE)

  output$plot <- renderPlotly({
    plot_ly(sd, x = ~Sepal.Length, y = ~Sepal.Width)
  })
}

shinyApp(ui, server)
```

 When working with shiny create the shared dataset in the server function or some things that follow might not work as expected.

One can also use the `data` method on the crosstalk object in reactive expressions, which allows accessing the Javascript selection where crosstalk is not directly supported, like below in a custom UI block. Note that the argument `withSelection` is set to TRUE in order to retrieve the selection state of the rows.

```
library(DT)
library(shiny)
library(crosstalk)

ui <- fluidPage(
  fluidRow(
```

```
    column(4, uiOutput("text")),
    column(8, DTOutput("table"))
  )
)

server <- function(input, output) {
  sd <- SharedData$new(cars)

  output$text <- renderUI({
    # get selected rows
    n_selected <- sd$data(withSelection = TRUE) %>%
      dplyr::filter(selected_ == TRUE) %>%
      nrow()

    h3(n_selected, "selected items")

  })

  output$table <- renderDT({
    datatable(sd)
  }, server = FALSE)
}

shinyApp(ui, server)
```

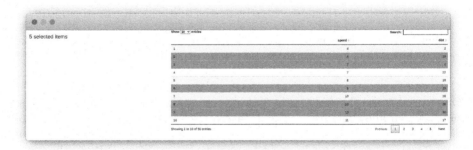

FIGURE 9.2: Shiny with crosstalk

Using crosstalk with shiny one can also change the selection server-side with
the `selection` method, passing it the keys to select.

```r
library(DT)
library(shiny)
library(crosstalk)

ui <- fluidPage(
  fluidRow(
    column(4, actionButton("random", "Select a random row")),
    column(8, DTOutput("table"))
  )
)

server <- function(input, output) {
  sd <- SharedData$new(cars)

  output$table <- renderDT({
    datatable(sd)
  }, server = FALSE)

  selected <- c()
  observeEvent(input$random, {
    smp <- c(1:50)[!1:50 %in% selected]
    selected <<- append(selected, sample(smp, 1))
    sd$selection(selected)
  })
}

shinyApp(ui, server)
```

9.2 Crosstalk Requirements

Crosstalk will not work well with every widget and every dataset. In some cases, it might not even be a good idea to support it at all.

Crosstalk works best on rectangular data: dataframes or objects that resemble dataframes like `tibble` or `SpatialPolygonsDataFrame`. This is important as crosstalk will treat the data row-wise, where each row is an observation that is ultimately selected, or filtered. If the underlying data is not tabular (e.g.: trees), then one might eventually encounter mismatches between widgets as they go out of sync.

Other than tabular data, crosstalk will require the widget to have the necessary functions or methods to dispatch the selection and filtering that crosstalk enables; that is, the widget must be able to filter as well as highlight and fade selected data points as crosstalk itself does not provide this.

9.3   How it Works

As will be discovered later when support for crosstalk is brought to gio, minimal changes of the R code is required. As might be expected, crosstalk enables the communication between widgets via JavaScript. Hence much of what must be adapted by widgets developers happens in JavaScript too as shown in Figure 9.3.

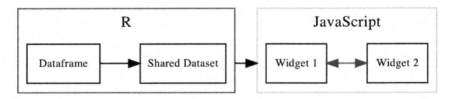

FIGURE 9.3: Crosstalk visualised

Indeed the bi-directional communication between widgets works in the viewer, , Shiny, and elsewhere, clearly indicating that all of it is taking place in the browser.

9.3.1   Keys

This internally works with keys that are assigned to every row of the dataframe, which enable crosstalk to track which are selected or filtered.

When creating shared datasets crosstalk will by default use the row names of the data.frame, and if these are not available, the SharedData function will create row numbers.

```
sd_cars <- SharedData$new(cars[1:2,])
```

You can therefore *mentally represent* the above-shared dataset as the following table. Note the emphasis; internally crosstalk does not actually add a column to the dataset: it leaves it as-is.

key	speed	dist
1	4	2
2	4	10
3	7	4
4	7	22
5	8	16

The keys assigned can be retrieve with the `key` method on the shared dataset itself.

```
sd_cars$key()
#> [1] "1" "2"
```

Otherwise these keys can be explicitly set by the user when creating the package.

```
# assign keys
df <- data.frame(x = runif(5))
sd <- SharedData$new(df, key = letters[1:5])
sd$key()
#> [1] "a" "b" "c" "d" "e"
```

9.3.2 Communication Lines

In a sense, while crosstalk establishes lines of communication to transport `keys` between widgets, developers of the respective widgets must handle what `keys` are sent to other widgets and what to do with incoming `keys` (that are selected or filtered in other widgets). There are two such lines of communication, one for `keys` of rows to be filtered, meant to narrow down the selection of data points displayed on a widget, and another for selection (what crosstalk refers to as "linked brushing") to highlight specific data points (fading out other data points).

In JavaScript, a widget "receives" the keys of selected and filtered data points and must, when filtering or selection is observed, "send" said selected or filtered keys to other widgets.

9.3.3 Groups

Internally crosstalk knows what to share across widgets; with `groups` that share `keys` and are isolated from each other so one can use multiple different shared datasets without them interfering with each other (see Figure 9.4).

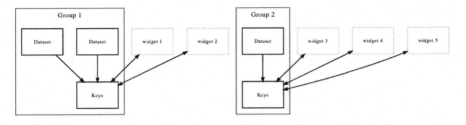

FIGURE 9.4: Crosstalk groups visualised

Crosstalk groups share keys.

Therefore, the code below creates two shared datasets that are linked and share keys as they fall in the same `group` even though they are separate R objects.

```
shared_cars <- SharedData$new(mtcars, group = "cars")
shared_cars_head <- SharedData$new(head(mtcars), group = "cars")
```

9.4 Crosstalk with Gio

The application of crosstalk to the gio library is somewhat amiss, but this makes it rather more instructive as it requires thinking beyond the mere implementation of the crosstalk and is an exercise the reader will likely have to do when incorporating it to other widgets. As mentioned before, in order for crosstalk to be properly implemented a widget must be able to select and deselect, as well as filter and unfilter data points, and this is not entirely the case of gio.

First, gio's underlying data is somewhat uncommon: it is a network defined only by its edges (the arcs leaving and coming into countries). Second, those edges themselves cannot be selected; as we've observed previously what edges are drawn on the globe cannot directly be defined; the selected country can be changed, which only by proxy changes the edges shown on the globe. Third, while gio supports changing which country is selected (by clicking on the globe), it does not allow unselecting a country; with gio.js a country is always selected.

The way crosstalk can work with gio is by setting the keys of the shared dataset to the country ISO codes that gio uses. Since data gio accepts consists of edges, this ISO code could correspond to either the source or the target country.

```r
# keys = target
shared_arcs <- SharedData$new(arcs, key = arcs$e)
# keys = source
shared_arcs <- SharedData$new(arcs, key = arcs$i)
```

This constraint would have to be documented and communicated to the users of the package as otherwise, gio's implementation of crosstalk will not work.

Were the constraint of having to specify the keys removed, the gio package would have to interpret the keys. For instance, the widget would receive a selection, say 3 indicating that the third edge was selected in another widget; gio, not being able to highlight the edge, would have to decide whether to highlight either the country where the edge comes from or the country where the edge goes to. Though this could be implemented, it would be vastly more laborious and be more limited as the choice of country to highlight would no longer up to the user.

9.5 R code

In any event, let us start by making the required changes to the R code first. The only changes that need to be made are in the gio function as it is the only one that currently accepts a dataframe and thus may receive a shared dataset.

```r
class(shared_arcs)
#> [1] "SharedData" "R6"
```

Shared datasets are R6 classes and therefore cannot simply be treated as dataframes. The gio function needs to check whether the data object it received is a shared dataset with `is.SharedData` and, if so, use its methods to extract data from it, namely:

- The original dataset with `origData`
- The group to which the dataset belongs with `groupName`
- The keys that were assigned to every row of the dataset with `key`

The `origData` method is needed to extract the original dataframe from the shared dataset. That is, of course, necessary as the `gio` function still needs to obtain and serialise the arcs to display on the globe.

```
# original data
shared_arcs$origData()
#>    e  i        v
#> 1 CN US 3300000
#> 2 CN RU   10000
```

The `gio` function also has to extract the group to which the dataset belongs; this will be necessary on the JavaScript-side to tell crosstalk which group one is working with. Note that it was randomly generated since none were specified when the shared dataset was created.

```
# groupName
shared_arcs$groupName()
#> [1] "SharedDatadf3b988c"

# keys
shared_arcs$key()
#> [1] "US" "RU"
```

The methods `origData` and `groupName` must be used in every widget, the `key` method may not be of use to every widget, it can be immensely useful if the visualisation library also comes with a key/id system so one can use it internally. Gio.js does not, and we thus will not be using it. The name of the group is passed to the `x` object, so it is accessible JavaScript-side where it is needed; we also add the JavaScript dependency required to run crosstalk with `crosstalkLibs`.

```r
gio <- function(data, width = NULL, height = NULL,
  elementId = NULL) {

  # defaults to NULL
  group <- NULL
  deps <- NULL

  # uses crosstalk
  if (crosstalk::is.SharedData(data)) {
    group <- data$groupName()
    data <- data$origData()
    deps <- crosstalk::crosstalkLibs()
  }

  # forward options using x
  x = list(
    data = data,
    style = "default",
    crosstalk = list(group = group) # pass group
  )

  attr(x, 'TOJSON_ARGS') <- list(dataframe = "rows")

  # create widget
  htmlwidgets::createWidget(
    name = 'gio',
    x,
    width = width,
    height = height,
    package = 'gio',
    elementId = elementId,
    sizingPolicy = htmlwidgets::sizingPolicy(
      padding = 0,
      browser.fill = TRUE,
      defaultWidth = "100%"
    ),
    preRenderHook = render_gio,
    # add crosstalk dependency
    dependencies = deps
  )
}
```

One could improve upon this section by using creating methods on the `gio` function. It would make for cleaner code, but this is outside the scope of this book.

9.6 JavaScript Code

What is left to do is to adapt the JavaScript code. As a reminder, it must accept the keys selected in other widgets and share the selected key with other widgets.

First, we create the selection handler in the `factory` function; this is done by instantiating a new class from `crosstalk.SelectionHandle`.

```
var sel_handle = new crosstalk.SelectionHandle();
```

Once the selection handle created it can be used in the `renderValue` function to set the group that was collected from R.

```
sel_handle.setGroup(x.crosstalk.group);
```

9.6.1 Send Selected Keys

In order for gio to share the selected country with other widgets, it would first have to know which country is selected. This can be achieved with a callback function that gio supports.

Most JavaScript visualisation libraries will support callbacks or events that are triggered when the user interacts with the visualisation so one can have arbitrary code run when, for example, a user clicks a point on a scatter plot, or when the user clicks the legend of a chart. What these callback functions will be and how they work will entirely depend on the library at hand.

In gio.js this callback function is fired when a country is selected on the globe, it accepts two objects: one containing data on the country selected and another containing data on the related countries (the arcs coming and leaving the selected country).

The documentation of gio.js[1] gives the following example callback function.

---

[1]https://giojs.org/html/docs/callbackPicked.html

```
// define callback function
function callback (selectedCountry, relatedCountries) {
  console.log(selectedCountry);
  // console.log(relatedCountries);
}


// use callback function
controller.onCountryPicked(callback);
```

```
{name: "LIBYA", lat: 25, lon: 17, center: n, ISOCode: "LY"}
```

This defines a function named callback, which takes the two objects as mentioned above and logs them in the JavaScript console. Then the function is passed to the controller via the onCountryPicked method, which will run it every time a country is selected by the user.

This callback function will be useful to send the keys to other widgets: when a user selects China to send the CN key selection via crosstalk.

As mentioned at the beginning of this section, the keys used with the datasets for gio.js should be country ISO codes. Therefore one can consider the variable selectedCountry.ISOCode as selected key. The set method from the selection handle can be used to share the selected key with other widgets. Note that this method expects either a null value or an array; a scalar value will throw an error, hence selectedCountry.ISOCode is wrapped in square brackets.

```
function callback (selectedCountry) {
  sel_handle.set([selectedCountry.ISOCode]);
}

controller.onCountryPicked(callback);
```

9.6.2 Set Selected Keys

We have implemented the necessary to share the selected country with other widgets but are yet to implement the opposite; when users select a country in another widget, the selected country in gio should change too. Gio does provide a method called switchCountry to change the selected country programmatically.

This can be achieved by listening to the change event on the selection handle previously created; below it is used to log the object `e` in order to inspect its properties.

```
// placed in factory function
sel_handle.on("change", function(e) {
  console.log(e);
});
```

```
{
  oldValue: [],
  sender: n {
    _eventRelay: e,
    _emitter: t,
    _group: "SharedDatac7682f87",
    _var: r,
    _varOnChangeSub: "sub1",
    …
  },
  value: ["AE"]
}
```

1. `oldValue` - the value that was previously selected (if any); this may be useful if the widget wants to calculate differences between the currently and previously selected value.
2. `sender` - the selection handle instance that made the change. This is useful to compare against the selection handle of the widget and know whether this widget or another initiated the selection change. It is often used to clear the selection or filtering before applying a new one when the change comes from another widget.
3. `value` - the array of selected keys.

Therefore event listener could make use of gio.js' `switchCountry`. Note that 1) the selection cannot be cleared with gio.js, a country is always selected, and 2) one can only select one country a time, hence only accepting the first element of the selected keys with `e.value[0]`.

```
// placed in factory function
sel_handle.on("change", function(e) {

  // selection comes from another widget
```

```
  if (e.sender !== sel_handle) {
    // clear the selection
    // not possible with gio.js
  }
  controller.switchCountry(e.value[0]);
});
```

9.7 Using Crosstalk with Gio

Finally, now that gio supports shared datasets, we can create a few examples to demonstrate how it can be used.

The simplest way is probably to convert the edges to a shared dataset specifying either the source (i) or target (e) country codes as keys. However, this is unlikely to be used this way out in the real world. In Figure 9.5, selecting an edge highlights a node, which is somewhat confusing.

```
library(DT)
library(gio)
library(crosstalk)

url <- paste0(
  "https://raw.githubusercontent.com/JohnCoene/",
  "javascript-for-r/master/data/countries.json"
)
arcs <- jsonlite::fromJSON(url)

# Wrap data frame in SharedData
# key is importing country
sd <- SharedData$new(arcs, key = arcs$i)

bscols(
  gio(sd),
  datatable(sd, width="100%", selection = "single")
)
```

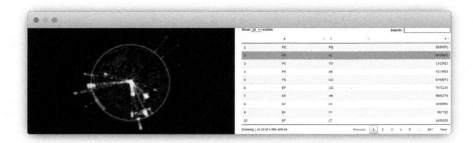

FIGURE 9.5: Gio with DT using crosstalk

Thankfully we can use the `group` argument in order to create edges and nodes that share keys (see Figure 9.6) and produce a more sensible link between widgets.

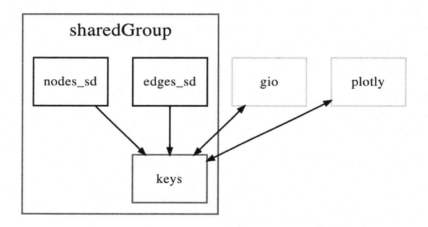

FIGURE 9.6: Crosstalk with gio

Below we create two shared datasets with the same group name, one for the edges and another for the nodes to produce Figure 9.7. Use one for the gio visualisation and the other for the plotly graph.

```
library(gio)
library(plotly)
library(crosstalk)
```

```r
url <- paste0(
  "https://raw.githubusercontent.com/JohnCoene/",
  "javascript-for-r/master/data/countries.json"
)
arcs <- jsonlite::fromJSON(url)

# Wrap data frame in SharedData
edges_sd <- SharedData$new(
  arcs, key = arcs$i, group = "sharedGroup"
)

# create nodes
iso2c <- unique(arcs$i)
nodes <- data.frame(
  country = iso2c,
  y = rnorm(length(iso2c))
)
nodes_sd <- SharedData$new(
  nodes, key = nodes$country,
  group = "sharedGroup"
)

bscols(
  plot_ly(data = nodes_sd, type = "bar", x = ~country, y = ~y) %>%
    config(displayModeBar = FALSE),
  gio(edges_sd)
)
```

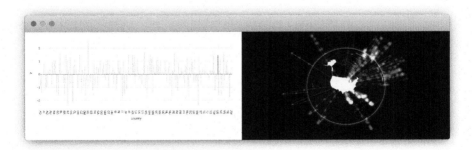

FIGURE 9.7: Gio and plotly using crosstalk and groups

10

Final Revisions

In this chapter, we polish the API that gio presents its users and provide guidelines to integrate other JavaScript libraries with R best.

10.1   Htmlwidgets and Data

The gio package built thus far revolves around the `gio` function, which expects a dataframe with three columns named `e`, `i`, and `v`, which is not great practice; there are ample reasons why very few functions do that.

First, it is unlikely that one comes across a dataset with such names in the real world thus users of the package will likely need to rename the columns of the dataset in order to use gio, making the package rather unwieldy. Second, this makes understanding and approaching the gio package more complicated; it will not be evident by looking at the examples, and usage of gio.

Instead `gio` should accept the dataframe as the first argument and then the relevant columns to extract. This can be implemented in many ways ranging from arguments that accept the column names as strings to reproducing ggplot2's `aes` function. Here we settle for using non-standard evaluation to provide arguments that accept the bare name of the columns.

```
gio <- function(data, source, target, value, ...,
  width = NULL, height = NULL, elementId = NULL) {

  # defaults to NULL
  group <- NULL

  if (crosstalk::is.SharedData(data)) {
    group <- data$groupName()
    data <- data$origData()
  }
```

```r
  data <- dplyr::select(
    data,
    i = {{ source }},
    e = {{ target }},
    v = {{ value }}
  )


  # forward options using x
  x = list(
    configs = list(...),
    data = data,
    style = "default",
    crosstalk = list(group = group)
  )


  attr(x, 'TOJSON_ARGS') <- list(dataframe = "rows")

  # create widget
  htmlwidgets::createWidget(
    name = 'gio',
    x,
    width = width,
    height = height,
    package = 'gio',
    elementId = elementId,
    sizingPolicy = htmlwidgets::sizingPolicy(
      padding = 0,
      browser.fill = TRUE,
      defaultWidth = "100%"
    ),
    preRenderHook = render_gio,
    dependencies = crosstalk::crosstalkLibs()
  )
}
```

The above changes allow documenting the input that gio accepts more clearly with roxygen2 and also makes its usage more transparent: it is now clear to users what data is required to create a visualisation, and they are free to use dataframes of their choice.

```r
# mock up data
countries <- c("US", "BE", "FR", "DE")
```

```r
df <- data.frame(
  from = countries,
  to = rev(countries),
  traded = runif(4)
)

# use gio
gio(df, source = from, target = to, value = traded)
```

This small change makes the package a great deal more comfortable to use and understand as source, and target are vastly more evident to understand than e and i.

10.2 Plethora of Options

Some JavaScript libraries can be extensive and come with thousands of options that can make the port to R rather bulky. Never hesitate to make use of the three dots construct (...) to make these accessible yet saving you from having to hard-code thousands of arguments.

For instance, gio.js accepts a JSON of options to customise the globe further. One could port all of these manually, or allow users to specify those configurations via the three-dot construct.

```javascript
var configs = {
  control: {
    stats: false,
    disableUnmentioned: false,
    lightenMentioned: false,
    inOnly: false,
    outOnly: false,
    initCountry: "CN",
    halo: true
  },
  color: {
    surface: 0xFFFFFF,
    selected: null,
    in: 0x154492,
    out: 0xDD380C,
```

```
    halo: 0xFFFFFF,
    background: null
  },
  brightness: {
    ocean: 0.5,
    mentioned: 0.5,
    related: 0.5
  }
}


controller = new Gio.controller(el, configs);
```

The three dots can be added to the `gio` function, which internally captures them in a `list` named `configs` so it can be easily referenced in JavaScript.

```
# add ...three dots
gio <- function(data, source, target, value, ...,
  width = NULL, height = NULL, elementId = NULL) {

  # ... start of the function

  # forward options using x
  x = list(
    configs = list(...), # pass to configs
    data = data,
    style = "default",
    crosstalk = list(group = group)
  )

  # ... end of the function
}
```

In JavaScript, use the `configs` when initialising the visualisation.

```
// use x.configs
controller = new GIO.Controller(el, x.configs);
```

Below those configuration options are now used to set the initially selected country to the United States and change the colour of the selected country to red in Figure 10.1.

```
gio(
  df, from, to, traded,
  control = list(initCountry = 'US'),
  color = list(selected = '#ff4d4d')
)
```

FIGURE 10.1: Gio and plenty of options

10.3   Interface Design

As you develop a wrapper to an external visualisation library, you will have to make design choices. In building gio, we more or less mirrored the JavaScript code one to one: where there is a JavaScript function to change the theme of the visualisation, there is one in R, etc. This might not scale appropriately as more and more functions are added to the package.

As observed, the gio.js library has a function named setStyle to change the theme of the visualisation, but it has numerous others, setSurfaceColor, addHalo, setHaloColor, removeHalo, and plenty more. We might want to wrap all or some of these in a single function to provide a more convenient API to the R user.

 Design for humans: always keep in mind the interface you make available to users as you develop the package.

You can always go beyond what the underlying library provides. For instance, the country selected by default is always China, regardless of whether the data includes that country or not. This can lead to creating underwhelming visualisations as no arcs appear. One can consider adding simple heiristics to the gio function to ensure that is not the case, or have the function throw a warning when the initial country is not present in the dataset.

Finally, consider R users' expectations. There are many prominent visualisation packages on CRAN already, users of the gio package will likely have used ggplot2 (Wickham et al., 2020a), plotly, or highcharter before. Though these provide somewhat different APIs, they set precedents. The more the API of gio resembles those, the easier it will be for new users to start using gio. However, do not let this restrict the package either. Never hesitate to do differently than ggplot2 if you think it will provide a better interface to your users.

10.4   Exercises

Widgets likely involve numerous concepts that are new to most readers. It is a good idea to try and work on a widget of your own to grasp the learnings of this part of the book entirely. A quick Google search for "JavaScript visualisation libraries" uncovers hundreds of candidate libraries that can be made accessible from R; below is but a small selection.

- chart.js[1] - simple yet flexible JavaScript charting
- cytoscape.js[2] - network theory library for visualisation and analysis
- Toast UI charts[3] - easy way to draw various and essential charts
- amcharts[4] - library for all your data visualization needs

---

[1] https://www.chartjs.org/
[2] https://js.cytoscape.org/
[3] https://ui.toast.com/tui-chart/
[4] https://www.amcharts.com/

Part III

Web Development with Shiny

11

Bidirectional Communication

Shiny is the web framework of choice for the R programming language. Since JavaScript and Shiny both run in web browsers it follows that they can run alongside one another as one can include JavaScript in such applications. However, often disregarded is the ability for Shiny's R server to communicate to the front end and vice versa. This collection of chapters aims to show precisely how this works. In this first part, we brush up on the essentials, so we understand how to include JavaScript in shiny applications.

Then again, the goal is not to write a lot of convoluted JavaScript. On the contrary, with little knowledge of the language the aim is to write as little as possible but demonstrate to the reader that it is often enough to vastly improve the user experience of Shiny applications.

11.1 WebSocket an Shiny

Shiny applications have two components: the user interface (UI) and the server function. These two components communicate via a WebSocket: a persistent connection that allows passing messages between the server and clients connecting to it. In the R server, this connection is managed by Shiny using the httpuv (Cheng and Chang, 2021) and WebSocket (Chang et al., 2021b) packages, while in clients connecting to the server this connection is managed with JavaScript, as depicted in 11.1.

DOI: 10.1201/9781003134046-11

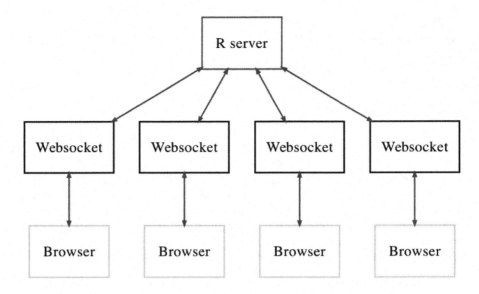

FIGURE 11.1: Websocket visualised

With that in mind, we can put together a Shiny application, which though simple. Exploits bi-directional communication. The application takes a text input, sends the value of the input to the R server, which sends it back to the UI.

```r
library(shiny)

ui <- fluidPage(
  textInput("nameInput", "Your name"),
  textOutput("nameOutput")
)

server <- function(input, output) {
  output$nameOutput <- renderText({
    input$nameInput
  })
}

shinyApp(ui, server)
```

Drawing a diagram of the communication between the UI and the server (Figure 11.2) reveals that though this is a simple application a lot is happening.

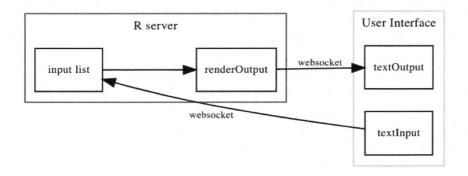

FIGURE 11.2: Shiny websocket visualised

Communicating between the R server and the user interface requires JavaScript and thus makes a reasonable chunk of this part of the book on web development with Shiny.

11.2 Sessions

Note that shiny isolate each client connecting to the server in what it refers to as "sessions." This means that when a user visits a shiny application and interacts with the inputs, like clicking a button, or moving a slider this only happens for their session, in their browser, and not in any other users'.

It would indeed be strange if when one of two concurrent users enters text in a box and that is reflected on the other user's screen.

This is good to know because WebSocket are in fact often use for precisely that effect. For instance, in a chat application where someone posting a message to a group chat is sent to the server which then, via the WebSocket, *broadcasts* the message to all other users in the group chat.

Shiny does not allow this, users are isolated from one another.

11.3   Alerts, an example

Let us exploit an external library to demonstrate how this works: jBox[1] allows displaying "notices," similar to vanilla JavaScript alerts but much better looking and with additional functionalities.

The grand scheme is to build an application that displays a notification at the click of an `actionButton` and "tells" the server when it is closed. Though the introduction of this book includes best practices on how to include dependencies and JavaScript files, much of that will be disregarded in this section (and only in this section). That is only so it does not get in the way of explaining bidirectional communication through the WebSocket in Shiny.

Moreover, the jBox library comes with numerous features to display tooltips, modals, notices, and more, which would make for too long a chapter; only notices shall be covered here. Let us first take a look at the code that generates a jBox notice.

11.3.1   Explore

Below we build an elementary example that features jBox in HTML; it includes the dependencies and a short script that displays a notification when the page is loaded.

```
<!DOCTYPE html>
<html xmlns="http://www.w3.org/1999/xhtml" lang="" xml:lang="">
<head>
<script
  src="https://code.jquery.com/jquery-3.5.1.min.js">
  </script>
<script
  src="https://cdn.jsdelivr.net/gh/StephanWagner/jBox@v1.2.0/
    dist/jBox.all.min.js">
</script>
<link
  href="https://cdn.jsdelivr.net/gh/StephanWagner/jBox@v1.2.0/
    dist/jBox.all.min.css"
  rel="stylesheet">
```

---

[1] https://github.com/StephanWagner/jBox

```
</head>

<body>
  <!-- Script to show a notification -->
  <script>
    new jBox('Notice', {
      content: 'Hurray! A notice!'
    });
  </script>
</body>
</html>
```

The very first thing one should do is recreate this basic example in Shiny so that when the app is loaded the notification appears, we will make this work with the bidirectional WebSocket communication afterwards. The "j" in jBox stands for jQuery, which is already a dependency of Shiny itself. There is, therefore, no need to import it; on the contrary one should not in order to avoid clashes.

```
library(shiny)

ui <- fluidPage(
  tags$head(
    tags$script(
      src = paste0(
        "https://cdn.jsdelivr.net/gh/StephanWagner/",
        "jBox@v1.2.0/dist/jBox.all.min.js"
      )
    ),
    tags$link(
      rel = "stylesheet",
      href = paste0(
        "https://cdn.jsdelivr.net/gh/StephanWagner/",
        "jBox@v1.2.0/dist/jBox.all.min.css"
      )
    )
  ),
  tags$script("
    new jBox('Notice', {
      content: 'Hurray! A notice!'
    });"
  )
)
```

```
server <- function(input, output) {}

shinyApp(ui, server)
```

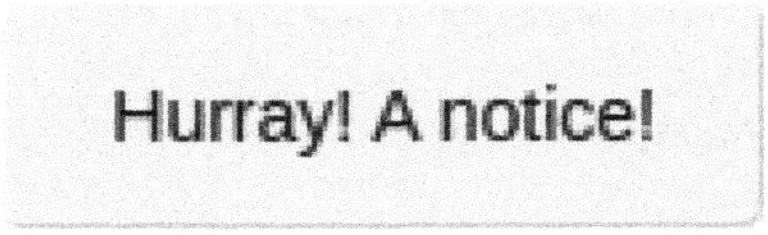

Hurray! A notice!

FIGURE 11.3: A basic jBox notice

Figure 11.3, and the application above essentially reproduce the basic HTML example that was shared; the dependencies are imported, and a script displays a notification. Since all of that takes place in the front end, the body of the server function is empty.

11.3.2   From R to JavaScript

Now that we have a simple notice displayed in the application we can tie it with the R server the alert should display a message sent by the R server. This would enable more dynamic messages, such as displaying a message taken from a database or a user input. As might be expected, there are two functions required to do so: an R function and its JavaScript complementary. One sends the data from the server and another catches said data in the client and displays the notice.

Let us start by writing the R code to send the data–thankfully very little is required of the developer. One can send data from the R server to the client from the `session` object using the `sendCustomMessage` method. Note that being a method of the `session` object implies that this message will only be sent to said session: only the client connected to that session will receive the message.

The `sendCustomMessage` method takes two arguments: first an identifier (`type`), second the actual data to send to JavaScript (`message`). The identifier passed first will be necessary JavaScript-side to "catch" that message and show the notice.

```r
server <- function(input, output, session){
  # set the identifier to send-notice
  session$sendCustomMessage(
    type = "send-notice", message = "Hi there!"
  )
}
```

So while the above sends the message to JavaScript through the WebSocket, nothing is yet to be done with that message once it arrives in the client. We can add a "handler" for the identifier we defined (`send-notice`) which will do something with the message we sent from the server. This is done with the `addCustomMessageHandler` method from the `shiny` object, where the first argument is the identifier and the second is the function that handles the message, a function that takes a single argument: the data sent from the server.

Below we add the handler for messages of type `send-notice`; the handler itself is a function that accepts the messages that were sent from the server and uses it to generate the notice via jBox.

```js
tags$script(
  "Shiny.addCustomMessageHandler(
    type = 'send-notice', function(message) {
      new jBox('Notice', {
        content: message
      });
  });"
)
```

This effectively enables passing a message that is taken from a database, for instance, or as shown below from a user input, to the front end, which generates a notice.

```r
library(shiny)

ui <- fluidPage(
  tags$head(
    tags$script(
```

```
      src = paste0(
        "https://cdn.jsdelivr.net/gh/StephanWagner/",
        "jBox@v1.2.0/dist/jBox.all.min.js"
      )
    ),
    tags$link(
      rel = "stylesheet",
      href = paste0(
        "https://cdn.jsdelivr.net/gh/StephanWagner/",
        "jBox@v1.2.0/dist/jBox.all.min.css"
      )
    )
  ),
  textInput("msg", "notice"),
  actionButton("notify", "Notify myself"),
  tags$script(
    "Shiny.addCustomMessageHandler(
      type = 'send-notice', function(message) {
        new jBox('Notice', {
          content: message
        });
    });"
  )
)

server <- function(input, output, session) {

  observeEvent(input$notify, {
    session$sendCustomMessage(
      type = "send-notice", message = input$msg
    )
  })

}

shinyApp(ui, server)
```

notice

Hello readers!

Notify myself

Hello readers!

FIGURE 11.4: A notice triggered by the server

In the previous application that produces Figure 11.5, notice the path that the message follows: it goes from the client (user input) to the server (observeEvent), which sends (sendCustomMessage) it back to the client.

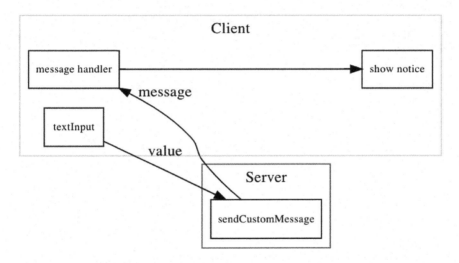

FIGURE 11.5: Shiny alert with custom messages

This might be considered suboptimal by some as it is not necessary to use the server as an intermediary (in this example at least). Though there is some truth to this, the above will work perfectly fine–and the aim here is to make JavaScript work with R–not alongside it. The WebSocket is very efficient, and this will not have much overhead at all.

11.3.3 Serialisation

Let us delve deeper into the data that is sent from the server to the front end to understand how we can further customise the notice displayed, e.g., change the colour.

```
new jBox('Notice', {
  content: 'Hurray! A notice!',
  color: 'red'
});
```

The jBox notice is configured using a JSON object containing the options that define said notice to display (example above), including but not limited to the message. The most straightforward way to make all those options accessible to the server is to construct that list of options server-side before sending it to the front end. For instance, the JSON of options displayed above would look like the R list below.

```
options <- list(
  content = 'Hurray! A notice!',
  color = 'red'
)
jsonlite::toJSON(options, pretty = TRUE, auto_unbox = TRUE)
#> {
#>   "content": "Hurray! A notice!",
#>   "color": "red"
#> }
```

Therefore, one could construct this list server-side and use it in jBox straight-away. Doing so means the JavaScript code can be simplified to `new jBox('Notice', message);` and produce Figure 11.6.

```
library(shiny)

ui <- fluidPage(
  tags$head(
    tags$script(
      src = paste0(
        "https://cdn.jsdelivr.net/gh/StephanWagner/",
        "jBox@v1.2.0/dist/jBox.all.min.js"
      )
```

```
    ),
    tags$link(
      rel = "stylesheet",
      href = paste0(
        "https://cdn.jsdelivr.net/gh/StephanWagner/",
        "jBox@v1.2.0/dist/jBox.all.min.css"
      )
    ),
    tags$script("Shiny.addCustomMessageHandler(
      type = 'send-notice', function(message) {
        // use notice send from the server
        new jBox('Notice', message);
    });")
  )
)

server <- function(input, output, session){
  # define notice options
  notice = list(
    content = 'Hello from the server',
    color = 'black'
  )
  # send the notice
  session$sendCustomMessage(
    type = "send-notice", message = notice
  )
}

shinyApp(ui, server)
```

FIGURE 11.6: Customised jBox notice

11.3.4   JavaScript to R

Thus far we have covered how to pass data from the R server to JavaScript in order to display a notification, but we have yet to make the data travel the other way: from JavaScript to R. In this example, we would like to send data from JavaScript to R when the notice is closed (either by itself or the user).

One ubiquitous way that such feedback is enabled in JavaScript is through events and callback functions, which are triggered upon an action being performed by the user (like the click of a button) or when other interesting things happen in the code. jBox provides numerous such events[2] so functions can be used when a modal is created or when a notice is closed, for instance.

The concept of the callback function is not totally foreign to R, albeit rarely used. For instance, Shiny comes with this feature, `shiny::onStop` and `shiny::onStart`. These allow having functions run when the application starts or exits, very useful to clean up and close database connections when the app exits.

```r
server <- function(input, output){
  shiny::onStop(
    # callback function fired when app is closed
    function(){
      cat("App has been closed")
    }
  )
}
```

In jBox, these callback functions are included in the JSON of options; below the `onClose` event is fired when the notice is closed.

```
{
  content: 'Alert!',
  onClose: function(){
    // Fired when closed
    console.log('Alert is closed');
  }
}
```

This raises one issue: one cannot truly serialise to executable code. The attempt below serialises the function to a string that *will not* be evaluated in JavaScript,

---

[2]https://stephanwagner.me/jBox/options#events

just like "`function(x){ x + 1 }`" is not evaluated in R: it is not a function, it is a string.

```
# try to serialise an R function
jsonlite::toJSON("function(x){x + 1}", auto_unbox = TRUE)
#> "function(x){x + 1}"
```

One solution is to append the callback function to the object of options JavaScript-side.

```
tags$script("Shiny.addCustomMessageHandler(
  type = 'send-alert', function(message) {
    // append callback
    message.onClose = function(){
      // TODO send data back to R
    }
    new jBox('Notice', message);
});")
```

Placing a function inside a JSON object is expected in JavaScript, in contrast with R where though it works it is rarely if ever done (reference class/R6 are somewhat similar). The above JavaScript code to append the callback function could look something like the snippet below in R.

```
message <- list(content = "hello")
message$onClose <- function(msg){
  print(msg)
}
message$onClose("Closing!")
#> [1] "Closing!"
```

This explains how the event is used in jBox (and many other libraries), but the body of the callback used previously is empty and therefore will not do anything: we need it to send data back to the R server so it can be notified when the notice is closed.

This can be done by defining a simplified Shiny input. While the book will eventually cover fully-fledged Shiny inputs that can be registered, updated, and more, there is also a simplified version of the latter, which allows sending reactive input values to the server, where it can be used just like any other inputs (`input$id`). The value of the input can be defined using the `setInputValue` method, which takes the `id` of the input and the `value` to give it.

Below place `Shiny.setInputValue('notice_close', true)` in the body of the function so the input `input$notice_close` will be set to TRUE when the notice closes.

```
tags$script("Shiny.addCustomMessageHandler(
  type = 'send-alert', function(message) {
  // append callback
  message.onClose = function(){
    Shiny.setInputValue('notice_close', true);
  }
  new jBox('Notice', message);
});")
```

However, Shiny internally optimises how those values are set. First, if the input is set to the same value, then Shiny ignores it. This is fine if you are interested in the actual value of the input but will not work as expected if the input is to be used as an event. Indeed if you want to use this input in an `observe`, `observeEvent`, or `eventReactive`, you want it to be triggered every time the input changes, regardless of whether that value is the same as before. The second optimisation Shiny does is when the input is set to multiple different values before these have been processed, then only the most recent value will actually be sent to the server.

One can opt-out of these optimisations using the `priority: "event"` option when setting the input value, which is what we shall do here. We are not interested in the actual value of that input TRUE; we want to make sure the server gets notified every time a notification closes and given the aforementioned optimisations, it will not. The first time the event will be fired the input will be set from NULL to TRUE, but every subsequent notification that close will not fire that event since the value will not change (it's always sending TRUE to the server).

```
tags$script("Shiny.addCustomMessageHandler(
  type = 'send-alert', function(message) {
  // append callback
  message.onClose = function(){
    Shiny.setInputValue(
      'notice_close', true, {priority: 'event'}
    );
  }
  new jBox('Notice', message);
});")
```

That done it can be incorporated into the application (Figure 11.7) built thus far. Something interesting could be done server-side, but to keep things brief and straightforward, we merely print the value of the input to the R console.

```r
library(shiny)

ui <- fluidPage(
  tags$head(
    tags$script(
      src = paste0(
        "https://cdn.jsdelivr.net/gh/StephanWagner/",
        "jBox@v1.2.0/dist/jBox.all.min.js"
      )
    ),
    tags$link(
      rel = "stylesheet",
      href = paste0(
        "https://cdn.jsdelivr.net/gh/StephanWagner/",
        "jBox@v1.2.0/dist/jBox.all.min.css"
      )
    ),
    tags$script(
      "Shiny.addCustomMessageHandler(
        type = 'send-notice', function(message) {
          message.onClose = function(){
            Shiny.setInputValue(
              'notice_close', true, {priority: 'event'}
            );
          }
          new jBox('Notice', message);
      });"
    )
  ),
  textInput("msg", "A message to show as notice"),
  actionButton("show", "Show the notice")
)

server <- function(input, output, session){

  observeEvent(input$show, {
    # define notice options
    notice = list(
      content = input$msg,
      color = 'black'
```

```
    )

  # send the notice
  session$sendCustomMessage(
    type = "send-notice", message = notice
  )
})

  # print the output of the notice_close event (when fired)
  observeEvent(input$notice_close, {
    print(input$notice_close)
  })
}

shinyApp(ui, server)
```

```
#> [1] TRUE
```

A message to show as notice

| Hello readers! |

| Show the notice |

Hello readers!

FIGURE 11.7: jBox final application

In the next chapter, we will build another application that makes use of
bidirectional communication but also introduces a few more concepts to improve
how such communication takes place and allows passing more complex messages
from R to JavaScript and vice versa.

12

A Complete Integration

Thus far, this part of the book has covered both ways data travels between JavaScript and R in Shiny. However, the notices displayed in the previous chapter, though they demonstrate how both languages can work together within Shiny, come short of illustrating some more advanced use cases, how to package such code and more.

We shall introduce a fascinating JavaScript library that enables running machine learning models in web browsers: ml5.js[1]. The library is a high-level interface to tensorflow.js[2] but very extensive as it includes a multitude of models to deal with sound, image, text, and more. In this chapter, one of those models is implemented, an image classifier using mobileNet[3] but the method shown can be used to integrate any other model later on.

This is not a gimmick; running a model this way means it runs in the client (web browsers) and not on the Shiny server, leaving it free to compute anything else and serve other concurrent users. It's also fast; JavaScript is often wrongly believed to be slow, on the contrary. Finally, the JavaScript API provided is straightforward; it's impressive how ml5.js exposes complex models through such a simple API.

For those who may already know TensorFlow and want to use a lower-level library, the genius of tensorflow.js[4] is that it runs on WebGL and is therefore GPU-accelerated; i.e., it's not slow, and has a very similar API to the TensorFlow Python library.

We start by exploring ml5.js, then plan the Shiny application that will make use of it, and finally, we wrap our work in the form of an R package.

---

[1] https://learn.ml5js.org/
[2] https://www.tensorflow.org/js
[3] https://arxiv.org/abs/1704.04861
[4] https://www.tensorflow.org/js

12.1   Discover

As for all projects that involve external libraries, the very first thing to do is to scan the documentation to understand how it is used. The documentation of ml5.js is exemplary, filled with examples and crystal clear. It gives the following example[5] for the image classifier.

```
// Initialize the Image Classifier method with MobileNet
const classifier = ml5.imageClassifier('MobileNet', modelLoaded);

// When the model is loaded
function modelLoaded() {
  console.log('Model Loaded!');
}

// Make a prediction with a selected image
classifier.classify(
  document.getElementById('image'), (err, results) => {
    console.log(results);
  }
);
```

First, the image `classifier` is initialised from the `ml5` object with the `imageClassifier` method. This method takes two arguments: the name of the pre-trained model to use (`MobileNet`) and a callback function that is run when the model is loaded. The `classify` method from the `classifier` is used with, again, two arguments: 1) the DOM element that contains the image (`<img />`) and 2) a callback function to do something with the results of the classification.

Now we can jump to the next section to think about how this can be implemented in Shiny.

12.2   Setup

In Shiny, a dropdown menu could be provided to choose from pre-selected images, and upon selection, the server renders the selected image. At the click

---

[5]https://learn.ml5js.org/#/reference/image-classifier

of a button the model then runs and sends the results to the R server, which prints them in the UI (see Figure 12.1).

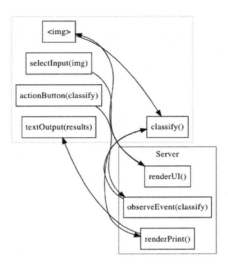

FIGURE 12.1: Simple shiny app, complex communication

This makes for what is probably a signature of Shiny: a considerable amount of bi-directional communication between the server and client as Shiny makes the most of the WebSocket. Some readers with more advanced knowledge in JavaScript will find ways to avoid the use of the server in places to do more in the client; either way works.

12.3 Dependencies

The ml5.js framework and all its components are bundled in a single JavaScript file.

```
<script src="https://unpkg.com/ml5@0.4.3/dist/ml5.min.js"></script>
```

We will create an `html_dependency` object using the `htmlDependency` function from the htmltools package. If confused, go back to the first part of the book on Shiny prerequisites, where it is explained in greater detail.

We have two options at our disposal either use the CDN (as shown in the previous code chunk) or download the file. We will start by making use of the CDN; later when we build a package for this functionality, we shall download it to provide users of the package a choice between using the local file or the CDN.

```
dependency_ml5 <- htmltools::htmlDependency(
  name = "ml5",
  version = "0.4.3",
  src = c(href = "https://unpkg.com/ml5@0.4.3/dist/"),
  script = "ml5.min.js"
)
```

12.4   Static Files

Images will, of course, be necessary in order to test the image classifier. We are therefore going to download some from Wikipedia. The following code chunk creates a directory of assets, downloads images of birds and saves them to the aforementioned directory. For brevity, we limit ourselves to downloading two images, one of a flamingo and another of a lorikeet, but feel free to add more. Also, note that the pre-trained image classifier we are going to use in this example is not limited to birds.

```
# static files directory
dir.create("assets")

# flamingo
fl <- paste0(
  "https://upload.wikimedia.org/wikipedia/",
  "commons/thumb/7/72/American_flamingo",
  "_%28Phoenicopterus_ruber%29.JPG/256px-",
  "American_flamingo_%28Phoenicopterus_ruber%29.JPG"
)

# lorikeet
lo <- paste0(
  "https://upload.wikimedia.org/wikipedia/",
  "commons/thumb/c/c2/Rainbow_lorikeet.jpg/",
  "256px-Rainbow_lorikeet.jpg"
```

```
)

# download
download.file(fl, destfile = "assets/flamingo.jpg")
download.file(lo, destfile = "assets/lorikeet.jpg")
```

Finally we should also add a JavaScript file, which will eventually contain our custom functions to run the image classifier.

```
file.create("www/classify.js")
```

At this stage, one should obtain a directory resembling the tree below.

```
.
├── app.R
└── assets
    ├── classify.js
    ├── flamingo.JPG
    └── lorikeet.jpg
```

These files will eventually need to be served (addResourcePath), so they are accessible by the Shiny UI.

12.5 Skeleton

At this stage, it's probably good to build a skeleton of the application (Figure 12.2).

After loading the Shiny package, we use the addResourcePath function to serve the images so they can be made accessible by the Shiny UI to display. At this stage, the application itself only provides a dropdown to select one of the two images previously downloaded, and a button to trigger the classification, which currently does not do anything, we'll delve into this next. Since we placed the classify.js JavaScript file in the assets directory we can also import it in the UI with a script tag; importantly this is done *after* the ml5.js dependency as it will depend on it. Another crucial thing that the app does is set the attribute id of the <img> to bird it is essential to have a convenient way to uniquely identify the image later on as ml5.js will need to read this image in order to classify it.

```r
library(shiny)

# serve images
addResourcePath("assets", "assets")

# ml5js dependency
dependency_ml5 <- htmltools::htmlDependency(
  name = "ml5",
  version = "0.4.3",
  src = c(href = "https://unpkg.com/ml5@0.4.3/dist/"),
  script = "ml5.min.js"
)

ui <- fluidPage(
  dependency_ml5,
  tags$head(
    tags$script(src = "assets/classify.js")
  ),
  selectInput(
    inputId = "selectedBird",
    label = "bird",
    choices = c("flamingo", "lorikeet")
  ),
  actionButton("classify", "Classify"),
  uiOutput("birdDisplay")
)

server <- function(input, output, session) {

  output$birdDisplay <- renderUI({
    path <- sprintf("assets/%s.jpg", input$selectedBird)
    tags$img(src = path, id = "bird")
  })

}

shinyApp(ui, server)
```

bird

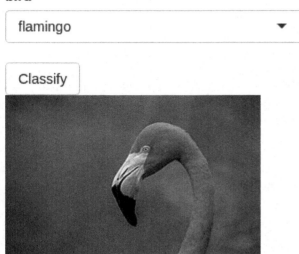

FIGURE 12.2: Shiny app skeleton

12.6 From R to JavaScript

What is now left to do is to program the classification. As a reminder, when the "classify" button is clicked, the classification must happen JavaScript-side using ml5.js; this implies that data must travel from R to JavaScript.

This will be carried in a similar fashion as in the previous chapter on alerts; the R server needs to send a message to the front end so it can trigger the classification using JavaScript.

```
observeEvent(input$classify, {
  session$sendCustomMessage("classify", list())
})
```

We thus observe the button so that when clicked, a message is sent to the front end, via the WebSocket. Note that the `sendCustomMessage` method **must take data,** hence the empty `list` that is used as the second argument. That, of course, won't do anything as we are yet to add a handler in JavaScript to handle this `classify` message that is sent.

Looking back at the documentation of ml5.js, we observe that before we can classify the image, the model should be loaded: we start by placing this code in the `classify.js` application.

The classifier is initialised from the `imageClassifier` method, which takes 1) the pre-trained model to use (or its name), and 2) a callback function. The callback function is run when the model is done loading. Though we don't make use of it here the argument is not optional (omitting it will raise an error) so we pass a function that simply prints `Model Loaded!` to the console.

```
// Mandatory callback function
function modelLoaded() {
  console.log('Model Loaded!');
}


// Initialize the Image Classifier method with MobileNet
const classifier = ml5.imageClassifier('MobileNet', modelLoaded);
```

There is no need to repeatedly initialise the classifier every time a user hits the "classify" button: this should only be done once.

Finally, we can take care of the message handler. Remember the message sent from the R server bears the `classify` unique identifier. The handler function runs the `classify` method on the previously instantiated `classifier` object. This takes 1) the image to classify and 2) a callback function to handle the results of the classification. Here we genuinely get to why we gave the generated `<img>` of the selected bird and `id`: it helps us quickly select that image from JavaScript to use in the classifier with `document.getElementById("bird")`.

```
// Mandatory callback function
function modelLoaded() {
  console.log('Model Loaded!');
}


// Initialize the Image Classifier method with MobileNet
const classifier = ml5.imageClassifier('MobileNet', modelLoaded);

Shiny.addCustomMessageHandler('classify', function(data){
  // Classify bird
  classifier.classify(
    document.getElementById("bird"), (err, results) => {
```

```
        console.log(results)
    }
  );
});
```

As mentioned at the start of the chapter, the results of the classification should be sent back to the R server, but for now, we shall content ourselves with logging it in the console.

Running the application and opening the console (Figure 12.3) already gives us encouraging results! The classifier gives "flamingo" the greatest confidence (albeit at 0.48).

FIGURE 12.3: Results logged to the console

12.7 From JavaScript to R

The application thus classifies the images, but the results remain in the front end, and we would like to have those results returned to the R server so we can further process them and display them back to the user.

As in the previous chapter, this can be done with the setInputValue function, which, as a reminder, will do exactly as advertised: it will set an input with a given value in the R server. The code below will make it such that the results will be accessible in the R server with input$classification.

```
// Mandatory callback function
function modelLoaded() {
  console.log('Model Loaded!');
}


// Initialize the Image Classifier method with MobileNet
const classifier = ml5.imageClassifier('MobileNet', modelLoaded);

Shiny.addCustomMessageHandler('classify', function(data){
  // Classify bird
  classifier.classify(
    document.getElementById("bird"), (err, results) => {
      Shiny.setInputValue("classification", results)
    }
  );
});
```

Now that the results are sent back to the R server, we can use them to display it back in the application (Figure 12.4) so users of the application may know how the model performed. We shall eventually make this prettier, but for now, we'll limit it to displaying the results in verbatimTextOutput.

```
library(shiny)

# serve images
addResourcePath("assets", "assets")

# ml5js dependency
dependency_ml5 <- htmltools::htmlDependency(
  name = "ml5",
  version = "0.4.3",
  src = c(href = "https://unpkg.com/ml5@0.4.3/dist/"),
  script = "ml5.min.js"
)

ui <- fluidPage(
  dependency_ml5,
```

```r
  tags$head(tags$script(src = "assets/classify.js")),
  selectInput(
    inputId = "selectedBird",
    label = "bird",
    choices = c("flamingo", "lorikeet")
  ),
  actionButton("classify", "Classify"),
  uiOutput("birdDisplay"),
  verbatimTextOutput("results") # display results
)

server <- function(input, output, session) {

  output$birdDisplay <- renderUI({
    path <- sprintf("assets/%s.jpg", input$selectedBird)
    tags$img(src = path, id = "bird")
  })

  observeEvent(input$classify, {
    session$sendCustomMessage("classify", list())
  })

  # render results
  output$results <- renderPrint({
    print(input$classification)
  })

}

shinyApp(ui, server)
```

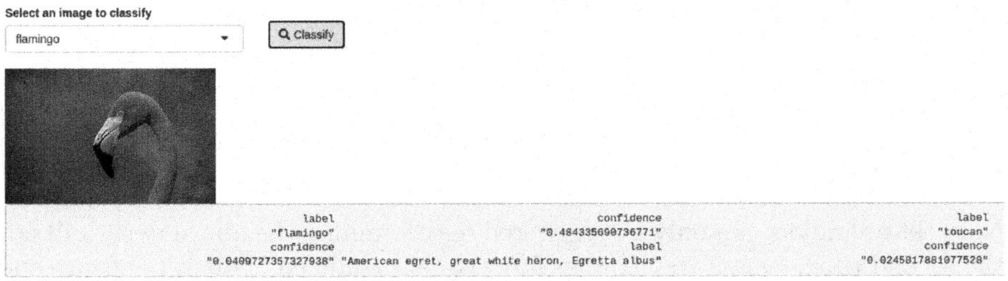

FIGURE 12.4: Classifier basic output

12.8   Input handler

In the previous section on sending data from R to JavaScript, we used a "message handler" in JavaScript to handle the data coming from the server. There is also the corollary, an "input handler" to preprocess the data coming from JavaScript before it is made accessible by the input. In R, this is a function that must accept three arguments: the data coming to JavaScript, a Shiny session, and the name of the input. Note that all of these arguments are mandatory if they are not used in the function we can use the three-dot construct instead.

Input handlers are most often used to reshape or change the type of the data coming in. To demonstrate how to use them, we will reshape the classification results sent to R as looking at the results of the classification in the R server one might notice a row-wise list, which can be transformed into a `data.frame`. The function below makes use of the purrr[6] (Henry and Wickham, 2020) package to loop over every result and transform them into data.frames and return a single data.frame.

```r
# create handler
process_results <- function(data, ...){
  purrr::map_dfr(data, as.data.frame)
}
```

Once this function created, it needs to be registered with Shiny using the `registerInputHandler` function, which takes two arguments. First, a unique identifier for the handler, second, the handler function. Attempt to give the handler a unique yet straightforward name (alphanumeric characters, underscores, and periods) to avoid clashes with other handlers.

```r
# register with shiny
shiny::registerInputHandler("ml5.class", process_results)
```

Note that handlers can only be registered once; running the above twice will fail the second time, even if the handler function has changed. This is to ensure one does not accidentally overwrite handlers brought in by other packages. These can be overwritten by explicitly setting `force` to `TRUE`, but it is not advised.

---

[6]https://github.com/tidyverse/purrr/

It is not advised to overwrite the registered handler.

```
# register with shiny
registerInputHandler("ml5.class", process_results)
```

Once the handler function is created and registered with Shiny, what is left to do is tell Shiny which input should use that handler. This is done by adding the name of the handler, ml5.class, preceded by a colon (:ml5.class) as a suffix to the input name.

```
Shiny.addCustomMessageHandler('classify', function(data){
  // Classify bird
  classifier.classify(
    document.getElementById("bird"), (err, results) => {
      Shiny.setInputValue("classification:ml5.class", results);
    }
  );
});
```

Now that the results of input$classification is a data.frame we can display the results in a neat table instead, as shown in Figure 12.5.

```
library(shiny)

# serve images
addResourcePath("assets", "assets")

# create handler
handler <- function(data, ...){
  purrr::map_dfr(data, as.data.frame)
}

# register with shiny
shiny::registerInputHandler("ml5.class", handler)

# ml5js dependency
dependency_ml5 <- htmltools::htmlDependency(
  name = "ml5",
  version = "0.4.3",
```

```
    src = c(href = "https://unpkg.com/ml5@0.4.3/dist/"),
    script = "ml5.min.js"
)

ui <- fluidPage(
  dependency_ml5,
  tags$head(tags$script(src = "assets/classify.js")),
  selectInput(
    inputId = "selectedBird",
    label = "bird",
    choices = c("flamingo", "lorikeet")
  ),
  actionButton("classify", "Classify"),
  uiOutput("birdDisplay"),
  tableOutput("results")
)

server <- function(input, output, session) {

  output$birdDisplay <- renderUI({
    path <- sprintf("assets/%s.jpg", input$selectedBird)
    tags$img(src = path, id = "bird")
  })

  observeEvent(input$classify, {
    session$sendCustomMessage("classify", list())
  })

  output$results <- renderTable({
    input$classification
  })

}

shinyApp(ui, server)
```

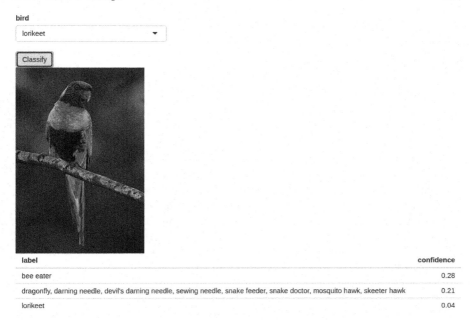

FIGURE 12.5: Classifier table output

12.9 As a Package

This chapter thus far built a nice application, but the code written is hardly portable; were one to make use of the image classifier from ml5.js in another application, everything would have to be rewritten or copy-pasted, which is hardly good practice and not remotely convenient. Instead this code should be packaged, so it is easily reusable and shareable. Moreover, this will benefit from all the other advantages that R packages bring to code such as documentation, reproducibility, and tests. This also forces the developer to think about the code differently. As we'll discover it's not as simple as wrapping individual functionalities from the app into functions.

Before we delve into building the package, let us think through what it should include. The application using ml5 gives some indication as to what the package will look like. Users of the package should be able to reproduce what is executed in the application, namely import dependencies (including the "message handler"), send data to the JavaScript front end to trigger the classification, and then obtain the results in the R server.

We start by creating a package called `ml5`.

```
usethis::create_package("ml5")
```

12.9.1 Dependencies

In the application, the web-hosted dependencies (CDN) were used. There are
two advantages to using CDNs: 1) it's just convenient as one does not have to
download them, 2) it's fast–CDNs are distributed geographically to improve
the speed at which they serve the dependencies and will therefore generally
outperform the alternative, serving the files locally. This may raise questions
when building a package though, as one generally wants these to be as modular,
self-contained, and reproducible as possible, and none of these things go well
with the idea of a remotely served dependency that is absolutely central to
the package. The package should therefore provide both ways of importing
dependencies: via the CDN or using locally-stored files. The former will be
faster while the latter can be used as a fallback in the event there is an issue
with the CDN or one does not have internet for instance.

We can download the dependency hosted on the CDN and place it in the `inst`
directory of the package. We also create another JavaScript `classify.js` that
will contain the custom JavaScript code (message handler, etc.) as was done
for the application.

```
# create directory
dir.create("inst")

# download dependency
uri <- "https://unpkg.com/ml5@0.4.3/dist/ml5.min.js"
download.file(uri, destfile = "inst/ml5.min.js")

# create js file
file.create("inst/classify.js")
```

With the dependencies locally downloaded one can move on to create the R
function that will be used to import the dependencies in the Shiny UI. The
file `classify.js` should be imported via this function too. The function `useMl5`
creates two `html_dependency` objects, one for the custom code with the message
handler and another for the ml5 dependency. Importantly at the end of the
function these are returned in a `tagList` **where order matters** as these will
be rendered in the order they are listed: first the ml5 dependency then the file
containing our custom code.

```r
# R/deps.R
#' @export
useMl5 <- function(cdn = TRUE) {

  # internal classify.js file
  pkg <- htmltools::htmlDependency(
    name = "ml5-pkg",
    version = "1.0.0",
    src = "",
    script = c(file = "classify.js"),
    package = "ml5"
  )

  # ml5 dependency
  if(cdn)
    ml5 <- htmltools::htmlDependency(
      name = "ml5",
      version = "0.4.3",
      src = c(href = "https://unpkg.com/ml5@0.4.3/dist/"),
      script = "ml5.min.js"
    )
  else
    ml5 <- htmltools::htmlDependency(
      name = "ml5",
      version = "0.4.3",
      src = "",
      script = c(file = "ml5.min.js"),
      package = "ml5"
    )

  htmltools::tagList(ml5, pkg)
}
```

12.9.2 Trigger classification

There will be a need for a function that sends a message to the front end to trigger the classification. In the application built previously the id of the image to classify was hard-coded this must be changed when building a package.

First, this will give users a much better interface where they may use whatever id suits them. Second, this will allow using the model to classify images that may be placed in different places and bear different ids.

```
# R/classify.R
#' @export
classify <- function(
  id,
  session = shiny::getDefaultReactiveDomain()
  ){
    session$sendCustomMessage("ml5-classify", id)
}
```

12.9.3   JavaScript code

As a quick reminder, the JavaScript should initialise the model and provide a
handler for the message `ml5-classify` that was defined in the previous section.
Nothing needs to change with regard to the initialisation of the model. However,
there are two things to adapt in the message handler. First, the `id` of the image
to classify is now dynamically defined and passed from the R server; the
code should therefore read `document.getElementById(data)` (where data is passed
from the server) instead of `document.getElementById('birds')` as was previously
hard-coded in the application.

Second, the application also had hardcoded the input id that was set with
the results of the classification (`input$classification`). This will no longer work
in a package: given the ability to classify multiple images the results of that
classification should set different inputs so as not to overwrite one another.
In the code below, we therefore create a dynamic input using the id: `id +
_classification`. Note that one can concatenate string in JavaScript using `+`,
while in R one would have to use the `paste0` function.

Examples:

- Classify   image   with   `id = 'birds'` will   return   results   to
  `input$birds_classification`
- Classify   image   with   `id = 'things'` will   return   results   to
  `input$things_classification`

```
// Initialize the Image Classifier method with MobileNet
const classifier = ml5.imageClassifier('MobileNet', modelLoaded);
// When the model is loaded
function modelLoaded() {
  console.log('Model Loaded!');
}
```

```
Shiny.addCustomMessageHandler('classify', function(data){
  // Classify bird
  classifier.classify(
    document.getElementById("bird"), (err, results) => {
      Shiny.setInputValue("classification:ml5.class", results);
    }
  );
});
```

12.9.4 Input handler

As mentioned the input handler that transforms the result sent from JavaScript
to R into a data.frame can only be registered once. Therefore, placing the code
that was written in an R file in the package will not work, or rather will work
only once.

```
# R/handler.R
handler <- function(data, ...){
  purrr::map_dfr(data, as.data.frame)
}

# This will error
# register with shiny
shiny::registerInputHandler("ml5.class", handler)
```

When the library is loaded the first time it will work, but all subsequent
attempts will fail.

```
library(ml5)

#> Loading ml5

library(ml5)

#> Loading ml5
#> Error in shiny::registerInputHandler("ml5.class", handler) :
#>   There is already an input handler for type: ml5.class
```

Packages can run functions when they are loaded or attached to an `.onLoad` function, which is called when the library is loaded in the global environment. The difference between loading and attaching a package can be subtle. In this case, it's probably best to run the function when the package is loaded using `.onLoad` which the R Packages book describes as:

Loading will load code, data, and any DLLs; register S3 and S4 methods; and run the `.onLoad()` function. After loading, the package is available in memory, but because it's not in the search path, you won't be able to access its components without using `::`. Confusingly, `::` will also load a package automatically if it isn't already loaded. It's rare to load a package explicitly, but you can do so with `requireNamespace()` or `loadNamespace()`.

— R Packages Book[7]

This could be used here to ensure the handler is only registered once; calling `library(packageName)` twice *does not* load the package twice, the second time it runs the function observes that the package is already loaded and thus does not load it again. The `.onLoad` replacement function may accept `libname` and `pkgname` arguments, we simplify the function to using `...` as we do not need those arguments.

```
# create handler
handler <- function(data, ...){
  purrr::map_dfr(data, as.data.frame)
}

# register with shiny
.onLoad <- function(...){
  shiny::registerInputHandler("ml5.class", handler)
}
```

12.9.5   Test

This is about all that is needed in order to build the package; one can then run `devtools::document()` to produce the man files and populate the namespace

---

[7]https://r-pkgs.org/

with the exported functions then install the package with `devtools::install()` in order to test it.

```r
library(DT)
library(ml5)
library(shiny)

addResourcePath("assets", "assets")

ui <- fluidPage(
  useMl5(),
  selectInput(
    inputId = "selectedBird",
    label = "bird",
    choices = c("flamingo", "lorikeet")
  ),
  actionButton("classify", "Classify"),
  uiOutput("birdDisplay"),
  DTOutput("results")
)

server <- function(input, output, session) {

  output$birdDisplay <- renderUI({
    path <- sprintf("assets/%s.jpg", input$selectedBird)
    tags$img(src = path, id = "bird")
  })

  observeEvent(input$classify, {
    classify("bird")
  })

  output$results <- renderDT({
    datatable(input$bird_classification)
  })

}

shinyApp(ui, server)
```

13

Tips and Tricks

While previous chapters on working with Shiny made use of external libraries and built packages that brought new functionalities previously not available in Shiny, one does not have to go to this length to take advantage of the learnings contained in those pages. Moreover, there are a few exciting things that have not yet been explored.

13.1 Shiny Events

There is a wee bit of documentation tucked away on the shiny website[1] that contains a useful list of events that Shiny fires to notify the developer of interesting things that happen in the application. This includes events that are fired when outputs are being recalculated, when Shiny connects, when an element become visible, and more. To demonstrate how to use those events and how handy they can be, we will create a notification which appears to indicate that the server is busy running computations. This could be as fancy as ever, but for simplicity's sake, we limit the demonstration to showing and hiding a gif.

First, we create the directories and necessary files, and to indicate the server is busy. We'll be using a gif that is rather well-known in the R community. Note that we will be using some CSS, hence the style.css file.

```r
dir.create("www")
file.create("www/script.js")
file.create("www/style.css")

# gif
gif <- paste0(
  "https://github.com/JohnCoene/javascript-for-r/",
```

---

[1]https://shiny.rstudio.com/articles/js-events.html

DOI: 10.1201/9781003134046-13

```
  "raw/master/code/events/www/typing.gif"
)
download.file(gif, "www/typing.gif")
```

Then we create an application that draws and redraws a plot at the click of a
button. Note that we give the gif an id as we will need to be able to retrieve
this element JavaScript-side (to dynamically show and hide it), and an id
makes for an ideal selector.

```
# app.R
library(shiny)

shiny::addResourcePath("www", "www")

ui <- fluidPage(
  # import dependencies
  tags$head(
    tags$link(href = "www/style.css", rel = "stylesheet"),
    tags$script(src = "www/script.js")
  ),
  # gif indicator
  tags$img(src = "www/typing.gif", id = "loading")
  plotOutput("plot"),
  actionButton("render", "render")
)

server <- function(input, output, session) {
  output$plot <- renderPlot({
    input$render # redraw on click

    # simulate time consuming computations
    Sys.sleep(2)
    plot(cars)
  })
}

shinyApp(ui, server)
```

The gif should only be visible when the server is busy, unlike now. Whether
it is visible will be controlled in JavaScript, so this should be initialised as
hidden using CSS. The following code hides the gif with `visibility: hidden`,
and repositions it, floating on top of the rest of the content in the top right of
the page, the `z-index` ensures the gif appears on top of other elements.

```css
/* style.css */
#loading{
  top: 20px;
  right: 20px;
  height: 200px;
  z-index: 9999;
  position: absolute;
  visibility: hidden;
}
```

We can then use the Shiny events to dynamically show and hide the gif when the server is busy. Below we observe the event shiny:busy on the entire page (document) when the event is triggered the gif is retrieved using its id and then made visible by changing its CSS visibility property to visible.

```javascript
// script.js
$(document).on('shiny:busy', function(event) {
  // retrieve the gif
  var title = document.getElementById("loading");

  // make it visible
  title.style.visibility = "visible";
});
```

We then need to hide the gif when the server goes from busy to idle, using the shiny:idle event we can change the visibility of the gif back to hidden.

```javascript
// script.js
$(document).on('shiny:busy', function(event) {
  // retrieve the gif
  var gif = document.getElementById("loading");

  // make gif visible
  gif.style.visibility = "visible";
});

$(document).on('shiny:idle', function(event) {
  var gif = document.getElementById("loading");

  // hide gif
  gif.style.visibility = "hidden";
});
```

The application will then display the gif when the server is busy running computations as in Figure 13.1.

FIGURE 13.1: Shiny with a busy indicator

13.2 Table Buttons

For instance, using what was learned previously, one can place buttons inside a Shiny table and observe server-side, which is clicked. With a basic application that only includes a table to which we ultimately want to add a column containing a button on each row. Here we achieve this by having each button set a different value (e.g., an id) to an input using `shiny.setInputValue`, but one could very well create different input names for each button.

```
library(DT)
library(shiny)

ui <- fluidPage(
  DTOutput("table")
)

server <- function(input, output) {

  output$table <- renderDT({
    datatable(
      mtcars,
      escape = FALSE,
      selection = "none",
      rownames = FALSE,
      style = "bootstrap"
    )
  })
}
```

```
}

shinyApp(ui, server)
```

Note that in the above we pass some parameters to datatable not all are necessary at the exception of escape, which is set to FALSE as we will ultimately place HTML code the table which should appear rendered rather than show said code as a string.

We start by creating the on-click functions as R character strings for each row of the mtcars dataset. This is the function that will be triggered when buttons are clicked. This should look familiar we use Shiny.setInputValue to define an input named click, which is set to a different value for every row of the table.

```r
library(DT)
library(shiny)

ui <- fluidPage(
  DTOutput("table")
)

server <- function(input, output) {

  output$table <- renderDT({
    # on click function
    onclick <- sprintf(
      "Shiny.setInputValue('click', '%s')",
      rownames(mtcars)
    )

    datatable(
      mtcars,
      escape = FALSE,
      selection = "none",
      rownames = FALSE,
      style = "bootstrap"
    )
  })

}

shinyApp(ui, server)
```

Next, we create the buttons for each row and set the JavaScript functions previously created as the onclick attributes. The JavaScript code passed to the onclick attribute will be executed every time the button is clicked.

```r
library(DT)
library(shiny)

ui <- fluidPage(
  DTOutput("table")
)

server <- function(input, output) {

  output$table <- renderDT({
    # on click function
    onclick <- sprintf(
      "Shiny.setInputValue('click', '%s')",
      rownames(mtcars)
    )

    # button with onClick function
    button <- sprintf(
      "<a class='btn btn-primary' onClick='%s'>Click me</a>",
      onclick
    )

    mtcars$button <- button
    datatable(
      mtcars,
      escape = FALSE,
      selection = "none",
      rownames = FALSE,
      style = "bootstrap"
    )
  })

}

shinyApp(ui, server)
```

We can then observe the click input and, to demonstrate, render it's value in the UI, see Figure 13.2 below.

```r
library(DT)
library(shiny)

ui <- fluidPage(
  br(),
  DTOutput("table"),
  strong("Clicked Model:"),
  verbatimTextOutput("model")
)

server <- function(input, output) {

  output$table <- renderDT({
    # on click function
    onclick <- sprintf(
      "Shiny.setInputValue('click', '%s')",
      rownames(mtcars)
    )

    # button with onClick function
    button <- sprintf(
      "<a class='btn btn-primary' onClick='%s'>Click me</a>",
      onclick
    )

    # add button to data.frame
    mtcars$button <- button

    datatable(
      mtcars,
      escape = FALSE,
      selection = "none",
      rownames = FALSE,
      style = "bootstrap"
    )
  })

  output$model <- renderPrint({
    print(input$click)
  })
}

shinyApp(ui, server)
```

Show 10 entries   Search:

mpg	cyl	disp	hp	drat	wt	qsec	vs	am	gear	carb	button
21	6	160	110	3.9	2.62	16.46	0	1	4	4	Click me
21	6	160	110	3.9	2.875	17.02	0	1	4	4	Click me
22.8	4	108	93	3.85	2.32	18.61	1	1	4	1	Click me
21.4	6	258	110	3.08	3.215	19.44	1	0	3	1	Click me
18.7	8	360	175	3.15	3.44	17.02	0	0	3	2	Click me
18.1	6	225	105	2.76	3.46	20.22	1	0	3	1	Click me
14.3	8	360	245	3.21	3.57	15.84	0	0	3	4	Click me
24.4	4	146.7	62	3.69	3.19	20	1	0	4	2	Click me
22.8	4	140.8	95	3.92	3.15	22.9	1	0	4	2	Click me
19.2	6	167.6	123	3.92	3.44	18.3	1	0	4	4	Click me

Showing 1 to 10 of 32 entries   Previous 1 2 3 4 Next

Clicked Model:

[1] "Mazda RX4"

FIGURE 13.2: DT with custom inputs

13.3   jQuery

The Shiny framework itself makes use of and thus imports the jQuery[2] JavaScript library, a library that provides a convenient API to make many things easier, including animations.

As an example, we could use jQuery's show, hide, or toggle functions to show or hide an HTML element at the press of a button.

```
// example of jQuery animation
$('#id').toggle();
```

Because jQuery is already imported, there is no need to do so, on the contrary, importing it again will impact load time and might clash with the pre-existing version. Below we create a Shiny application containing a message handler to toggle (show or hide element depending on its state) at the click of a button.

---

[2]https://jquery.com/

```r
library(shiny)

ui <- fluidPage(
  tags$head(
    tags$script(
      "Shiny.addCustomMessageHandler(
        'jquery-toggle', function(id){
          $('#' + id).toggle(); // id
      });"
    )
  ),
  actionButton("toggle", "Toggle text"),
  h1("This text is shown!", id = "text")
)

server <- function(input, output, session){

  observeEvent(input$toggle, {
    session$sendCustomMessage('jquery-toggle', "text")
  })

}

shinyApp(ui, server)
```

Note that jQuery takes a selector so one could very well use a class to hide
and show multiple elements (with said class) at once.

```r
library(shiny)

ui <- fluidPage(
  tags$head(
    tags$script(
      "Shiny.addCustomMessageHandler(
        'jquery-toggle', function(selector){
          $(selector).toggle();
      });"
    )
  ),
  actionButton("toggle", "Toggle text"),
  h1("This text is shown!", class = "to-toggle"),
  actionButton(
```

```
    "btn", "Another visible button", class = "to-toggle"
  )
)

server <- function(input, output, session){

  observeEvent(input$toggle, {
    session$sendCustomMessage('jquery-toggle', ".to-toggle")
  })

}

shinyApp(ui, server)
```

This is something where, again, R is leveraged in order to make it easier on the Shiny developer, but it must be said that it suffers from some inefficiency: the message travels from the browser (button click) to the R server, where it is sent back to the browser and triggers toggle. It could indeed very well be rewritten in JavaScript entirely. This is, however, outside the scope of this book.

14

Custom Outputs

In this chapter, we create a custom Shiny output; in practical terms, this creates custom `render*` and `*Output` functions to use in Shiny. This will be demonstrated by creating something akin to the `valueBox` available in the shinydashboard (Chang and Borges Ribeiro, 2018) package. While similar to what shinydashboard provides, this box will 1) be fully customisable and 2) available in any Shiny application 3) have additional functionalities.

The `valueBox` equivalent we shall build in this chapter is named "boxxy," and allows creating simple but colourful value boxes with animated numbers (by counting up to it) using countUp.js[1].

Figure 14.1 demonstrates what will be built in this chapter.

```r
library(shiny)

ui <- fluidPage(
  boxxyOutput("countries")
)

server <- function(input, output){
  output$countries <- renderBoxxy({
    boxxy("Countries", 95)
  })
}

shinyApp(ui, server)
```

---

[1]https://github.com/inorganik/countUp.js

FIGURE 14.1: Custom output example

14.1   Inner-workings

At the core, this consists in creating three functions: `boxxy`, `renderBoxxy`, and `boxxyOutput` (analogous to `plot`, `renderPlot`, `plotOutput`), which are linked by an "output binding" in JavaScript.

The first function, `boxxy` will accept arguments that help define what is in the box. This function is generally useful to preprocess any of the arguments that are meant to produce the custom output. The `boxxyOutput` function essentially creates the scaffold of the HTML output (e.g.: `<div>`), as well as the dependencies. The `render*` function is perhaps more peculiar: it should accept an expression and return a function.

Previous work with Shiny and JavaScript covered in this book had no dedicated "output" elements that were placed in the Shiny UI. Therefore, one had to use a function solely dedicated to importing the dependencies (e.g.: `usejBox`). However, since this is not the case here, the dependencies can be attached together with the output.

Finally, the two R functions are "bound" JavaScript-side with an "output binding" that renders the data from the `render*` function with its `*output`.

14.2 Setup

The custom output will be part of a Shiny application, let us thus create the basic skeleton of an application and download the dependencies. Create a project in RStudio or an empty directory, then:

1. Create an `app.R` file that will hold the code for the application and `boxxy`, `boxxyOutput`, and `renderBoxxy` functions.
2. Create an `assets` directory that will contain the CSS and JavaScript files.
3. Download the countUp.js dependency.
4. Create a `binding.js` JavaScript file for the JavaScript binding within the previously created `assets` directory.
5. Create a `styles.css` file in the `assets` directory.

```r
# application file
file.create("app.R")

# static file directory
dir.create("assets")

# countup dependency
url <- paste0(
  "https://cdn.jsdelivr.net/npm/",
  "countup@1.8.2/countUp.js"
)

download.file(url, "assets/countup.js")

# create binding file
file.create("assets/binding.js")

# CSS file
file.create("assets/styles.css")
```

This should produce the following directory structure.

```
.
├── app.R
└── assets
```

```
├── binding.js
├── countup.js
└── styles.css
```

14.3   Output R Function

The boxxy function takes three arguments: a title, a value that will be animated, and the background color to use for the box. The function, at this stage at least, does not preprocess the arguments and returns them as a named list.

```r
# app.R
library(shiny)

boxxy <- function(title, value, color = "#ef476f"){
  list(title = title, value = value, color = color)
}
```

14.4   Generate Output HTML

The boxxyOutput function, like all such functions (plotOutput, uiOutput, etc.) takes an id. This function should return an HTML tag that bears an id, or a data-input-id attribute (more on that later) and a class. The id is to be defined by the user of the function in Shiny just like any other such outputs. For instance, plotOutput creates a <div>, the id of which is actually the id used in the plotOutput function.

```r
# the id is used as id to the element
shiny::plotOutput(id = "myPlotId")
```

```html
<div
  id="myPlotId"
  class="shiny-plot-output"
  style="width: 100% ; height: 400px">
</div>
```

The `class` is used JavaScript-side to "find" the outputs in the DOM and generate the output. The function `boxxyOutput` could thus be as shown, the `id` is passed along to the `<div>`, which is created with a `boxxy` class that will be used in the output binding to find that element and generate the output within that very `<div>` using data that will be passed from the server.

```
boxxyOutput <- function(id){
  # the HTML output
  shiny::tags$div(
    id = id, class = "boxxy"
  )
}
```

Make sure you use unique class names so they are not accidentally overridden by the user.

As shown, the box should include a title and an animated value. These could be generated entirely in JavaScript, but it's actually easier to create placeholders with htmltools tags. We generate dynamic ids for those so they can easily be referenced later on in JavaScript: `id-boxxy-value` for the value and `id-boxxy-title` for the title.

```
boxxyOutput <- function(id){
  # the HTML output
  shiny::tags$div(
    id = id, class = "boxxy",
    h1(
      id = sprintf("%s-boxxy-value", id),
      class = "boxxy-value"
    ),
    p(
      id = sprintf("%s-boxxy-title", id),
      class = "boxxy-title"
    )
  )
}
```

Finally, we also used classes in the output so every element it comprises can be styled with ease.

```css
.boxxy{
  text-align: center;
  border-left: 6px solid #073b4c;
  padding: 1em;
}

.boxxy-title{
  text-transform: uppercase;
}

.boxxy-value{
  font-size: 3em;
}
```

In some previous examples we created a function decicated to importing dependencies in the Shiny UI but in this case they can piggyback on the boxxyOutput function. This works using the htmltools package. The function htmltools::htmlDependency is used to create a dependency that is then attached with htmltools::attachDependencies. While the former creates an object that Shiny can understand and translate into <script> or <style> tags, the former attaches them to the output object and ensures dependencies are not imported multiple times (e.g.: when boxxyOutput is used more than once).

Notice the use of normalizePath to retrieve the full path to the assets directory as this will not work with a relative path (e.g.: ./assets). The dependencies consist of the three files previously created and necessary to generate the output: countup.js, the dependency that was downloaded, as well as binding.js and styles.css.

```r
boxxyOutput <- function(id){
  el <- tags$div(
    id = id, class = "boxxy",
    h1(
      id = sprintf("%s-boxxy-counter", id),
      class = "boxxy-value"
    ),
    p(
      id = sprintf("%s-boxxy-title", id),
      class = "boxxy-title"
    )
  )

  # get full path
```

```
  path <- normalizePath("assets")

  deps <- list(
    htmltools::htmlDependency(
      name = "boxxy",
      version = "1.0.0",
      src = c(file = path),
      script = c("countup.js", "binding.js"),
      stylesheet = "styles.css"
    )
  )

  htmltools::attachDependencies(el, deps)
}
```

Running the function reveals the HTML it generates at the exception of the
dependencies which htmltools does not print to the console.

```
boxxyOutput("myID")
```

```
<div id="myID" class="boxxy">
  <h1 id="myID-boxxy-counter" class="boxxy-value"></h1>
  <p id="myID-boxxy-title" class="boxxy-title"></p>
</div>
```

14.5   Output Renderer

The function renderBoxxy should accept an expression, like other such render*
functions. For instance, in the example below the renderPlot function does
accept an expression that uses, amongst other functions, plot.

```
output$myPlot <- renderPlot({
  # this is an expression
  cars %>%
    head() %>%
```

```
    plot()
})
```

The function `renderBoxxy` takes an expression and other arguments that are passed to `shiny::exprToFunction`. This does pretty much what it says on the tin: it returns a function from an expression (unless that expression is a function, in which case it returns the expression). This function must be further wrapped in another as render functions must return functions.

```
renderBoxxy <- function(expr, env = parent.frame(),
  quoted = FALSE) {
  # Convert the expression + environment into a function
  func <- shiny::exprToFunction(expr, env, quoted)

  function(){
    func()
  }
}
```

14.6   JavaScript Output Binding

Here we create an "output binding." It tells Shiny how to find the component and how to interact with it. An output binding is initialised from `Shiny.OutputBinding`. Below we initialise a new binding.

```
// custom.js
var boxxyBinding = new Shiny.OutputBinding();
```

Then, this must be "extended" by specifying a number of methods, an essential one being `find`. It is used to look for the output HTML element in the document (`scope`), and to return them as an array (`HTMLcollection`). Other methods all take an `el` argument; that value will always be an element that was returned from `find`. A very straightforward way to accomplish this is to use jQuery's find method to identify elements with the `boxxy` class used in `boxxyOutput`. You are by no means forced to use a CSS class to identify the elements, but there is no reason not to.

```
// custom.js
var boxxyBinding = new Shiny.OutputBinding();

$.extend(boxxyBinding, {
  find: function(scope) {
    return $(scope).find(".boxxy");
  }
});
```

One might then want to use the `getId` method, which returns the `id` of the element, by default, as can be seen in the source code[2] (below), the binding returns the id as the `data-input-id` attribute, and if that is false it returns the element's `id`.

```
// getId default
this.getId = function(el) {
  return el['data-input-id'] || el.id;
}
```

Since boxxy uses the element id, the default will work, and this can be skipped entirely.

Next, one needs to implement the `renderValue` function, which is the same function that generates the output based on data used in `boxxy` and sent to the front end with `renderBoxxy`. The `renderValue` method accepts two arguments: first `el`, the element where the output should be generated; this is effectively the output of `boxxyOutput`, which the binding found using `find`. The second argument is `data` which is the data, passed to `boxxy` and serialised via `renderBoxxy`.

The `renderValue` is in effect very similar if not identical to the JavaScript function of the same name involved in creating htmlwidgets.

14.6.1 Boxxy Title

Let us now tackle the first of the three core aspects of the boxxy output: the title. The `title` should be placed in the previously-created placeholder which bears the `id-boxxy-title`, precisely as was done with htmlwidgets previously. We insert title (`data.title`) in the element with `innerText`. The dynamically

---

[2]https://github.com/rstudio/shiny/blob/master/srcjs/output_binding.js

generated id for the title is built in the same way it is in R, by concatenating
the `id` with `-boxxy-title`

- In R `sprintf("%s-boxxy-title", id)`
- In JavaScript `el.id + '-boxxy-title'`

```
var boxxyBinding = new Shiny.OutputBinding();

$.extend(boxxyBinding, {
  find: function(scope) {
    return $(scope).find(".boxxy");
  },
  renderValue: function(el, data) {

    // insert the title
    let title_id = el.id + '-boxxy-title';
    document.getElementById(title_id).innerText = data.title
  }
});
```

14.6.2   Boxxy Value

Though the custom output could be limited to a static value generated in a
fashion similar to how the title is placed, we opted for a more fancy animated
value using countUp.js.

Initialise a new counter, specify the id of the element, where it should be
created as first argument, as second argument the starting value from which
the counter should start, and finally the value to count up to. Note that there
is a fourth argument to pass a JSON of options, which we do not use here.

```
// place counter in elementId
// start at 0 and count up to 123
const counter = new CountUp('elementId', 0, 123);
counter.start();
```

The counter has to be generated in the `<h1>` placeholder bearing the `id-boxxy-value`, while the value to count up to is stored in `data.value` meaning the counter
can be initialised with `new CountUp(el.id + '-boxxy-value', 0, data.value);`.

```
var boxxyBinding = new Shiny.OutputBinding();

$.extend(boxxyBinding, {
  find: function(scope) {
    return $(scope).find(".boxxy");
  },
  renderValue: function(el, data) {

    // insert the title
    let title_id = el.id + '-boxxy-title';
    document.getElementById(title_id).innerText = data.title

    // counter start at 0
    let counter_id = el.id + '-boxxy-value';
    var counter = new CountUp(counter_id, 0, data.value);
    counter.start();
  }
});
```

14.6.3 Boxxy Background Color

Then we can set the background colour that was passed to boxxy (data.color).

```
var boxxyBinding = new Shiny.OutputBinding();

$.extend(boxxyBinding, {
  find: function(scope) {
    return $(scope).find(".boxxy");
  },
  renderValue: function(el, data) {

    // insert the title
    let title_id = el.id + '-boxxy-title';
    document.getElementById(title_id).innerText = data.title

    // counter start at 0
    let counter_id = el.id + '-boxxy-value';
    var counter = new CountUp(counter_id, 0, data.value);
    counter.start();

    // background color
```

```
      el.style.backgroundColor = data.color;
  }
});
```

14.6.4   Register the Output Binding

Finally, the output binding must be registered with Shiny. Note that it uses a unique string identifier. The documentation[3] states:

> At the moment it is unused but future features may depend on it.

```
var boxxyBinding = new Shiny.OutputBinding();

$.extend(boxxyBinding, {
  find: function(scope) {
    return $(scope).find(".boxxy");
  },
  renderValue: function(el, data) {

    // insert the title
    let title_id = el.id + '-boxxy-title';
    document.getElementById(title_id).innerText = data.title

    // counter start at 0
    let counter_id = el.id + '-boxxy-value';
    var counter = new CountUp(counter_id, 0, data.value);
    counter.start();

    // background color
    el.style.backgroundColor = data.color;
  }
});
```

---

[3]https://shiny.rstudio.com/articles/building-outputs.html

```
// register
Shiny.outputBindings.register(boxxyBinding, "john.boxxy");
```

Ensure that string uniquely identifies the binding to avoid future clash with other such bindings.

14.7   Boxxy Usage

With all of this in place, one can use boxxy in a Shiny application (Figure 14.2).

```
library(shiny)

boxxy <- function(title, value, color = "black"){
  list(title = title, value = value, color = color)
}

boxxyOutput <- function(id){
  el <- shiny::tags$div(
    id = id, class = "boxxy",
    h1(id = sprintf("%s-boxxy-value", id), class = "boxxy-value"),
    p(id = sprintf("%s-boxxy-title", id), class = "boxxy-title")
  )

  path <- normalizePath("assets")

  deps <- list(
    htmltools::htmlDependency(
      name = "boxxy",
      version = "1.0.0",
      src = c(file = path),
      script = c("countup.js", "binding.js"),
      stylesheet = "styles.css"
    )
  )

  htmltools::attachDependencies(el, deps)
```

```
}

renderBoxxy <- function(expr, env = parent.frame(),
  quoted = FALSE) {
  # Convert the expression + environment into a function
  func <- shiny::exprToFunction(expr, env, quoted)

  function(){
    func()
  }
}

ui <- fluidPage(
  h2("Custom outputs"),
  fluidRow(
    column(
      3, boxxyOutput("countries")
    ),
    column(
      3, boxxyOutput("employees")
    ),
    column(
      3, boxxyOutput("customers")
    ),
    column(
      3, boxxyOutput("subs")
    )
  )
)

server <- function(input, output){
  output$countries <- renderBoxxy({
    boxxy("Countries", 95, color = "#ef476f")
  })

  output$employees <- renderBoxxy({
    boxxy("Thing", 650, color = "#06d6a0")
  })

  output$customers <- renderBoxxy({
    boxxy("Customers", 13592, color = "#118ab2")
  })

  output$subs <- renderBoxxy({
```

```
    boxxy("Subscriptions", 16719, color = "#ffd166")
  })
}

shinyApp(ui, server)
```

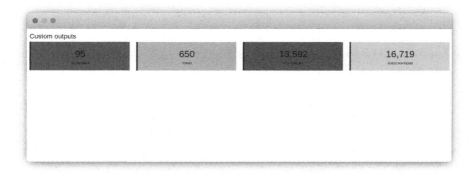

FIGURE 14.2: Shiny application with boxxy

14.8   Injecting Dependencies

We could consider making the animation of the value rendered with boxxy optional; some users may not want to use it. You might already imagine how a new argument and a few if-statements could very quickly do the job, but how would one handle the dependency? Indeed if users do not want to make use of the animation, the CountUp.js dependency should also be excluded so as to keep the output as light as possible.

The dependency is currently attached in the boxxyOutput function, which does not take any argument. It could, but it would make for the rather messy and confusing interface as whatever additional argument that indicates whether the numbers should be animated would have to be specified twice. Once in the boxxyOutput function, so it does not import the dependency, as well as in the boxxy function in order to serialise that parameter, so the JavaScript binding does not run the animation function.

```
# pseudo code
# do not do this
```

```r
library(shiny)

ui <- fluidPage(
  boxxyOutput(
    "countries",
    # do not import dependency
    animate = FALSE
  )
)

server <- function(input, output){
  output$countries <- renderBoxxy({
    # do not animate the numbers
    boxxy("Countries", 95, animate = FALSE)
  })
}

shinyApp(ui, server)
```

Thankfully there is a better way, combining htmltools and Shiny to insert the dependency dynamically from JavaScript.

The `boxxy` function needs to take an additional argument `animate`, which is passed to the output list. This will be used in the `render` function (and JavaScript binding) to render the dependency dynamically.

```r
boxxy <- function(title, value, color = "black", animate = TRUE){
  list(
    title = title, value = value, color = color, animate = animate
  )
}
```

The `boxxyOutput` function can be slightly simplified. It currently attaches the `countup.js` dependency, which needs to be removed.

```r
boxxyOutput <- function(id){
  el <- shiny::tags$div(
    id = id, class = "boxxy",
    h1(id = sprintf("%s-boxxy-value", id), class = "boxxy-value"),
    p(id = sprintf("%s-boxxy-title", id), class = "boxxy-title")
  )
```

```
  path <- normalizePath("assets")

  # only attach binding.js
  deps <- list(
    htmltools::htmlDependency(
      name = "boxxy",
      version = "1.0.0",
      src = c(file = path),
      script = c("binding.js"),
      stylesheet = "styles.css"
    )
  )

  htmltools::attachDependencies(el, deps)
}
```

The renderBoxxy function sees some modifications, while before it was technically only returning a function that itself returned the output of boxxy (func() == boxxy()). Here we want to capture the output of boxxy to check whether the animate element is TRUE and, if so, add the dependency.

```
renderBoxxy <- function(expr, env = parent.frame(),
  quoted = FALSE) {
  # Convert the expression + environment into a function
  func <- shiny::exprToFunction(expr, env, quoted)

  function(){
    val <- func()

    if(val$animate){
      # add dependency
    }

    return(val)
  }
}
```

Within the if statement, the dependency can be created with the htmltools as done for the binding. Ensure the names of the dependencies are unique as shiny internally uses it to differentiate between them; if they bear the same name Shiny assumes they are the same and will only render one of them.

Make sure dependencies bear different names or Shiny thinks it's the same and only renders one of them.

The dependency generated with htmltools is then passed to the `shiny::createWebDependency` function, which internally uses `shiny::addResourcePath` to serve the dependency. This is necessary here as, at this stage, the countup dependency is not actually rendered; below we merely add it to the list of options that serialised to JSON. Indeed, this will actually be injected JavaScript-side. Therefore the front end needs to be able to access this file, hence it is served.

```r
renderBoxxy <- function(expr, env = parent.frame(),
  quoted = FALSE) {
  # Convert the expression + environment into a function
  func <- shiny::exprToFunction(expr, env, quoted)

  function(){
    # evaluate to output list
    val <- func()

    # add dependency
    if(val$animate){
      path <- normalizePath("assets")

      deps <- htmltools::htmlDependency(
        name = "countup", # change name
        version = "1.8.2",
        src = c(file = path),
        script = c("countup.js") # only countup
      )

      # serve dependency
      val$deps <- list(shiny::createWebDependency(deps))
    }

    return(val)
  }
}
```

Thus far, the dependency is dynamically included in the R object; that is serialised to JSON, but it is not yet actually imported in the document–this happens in the JavaScript binding.

The first thing we ought to do is mirror the if-statement that was created in the `renderBoxxy` function. If the numbers should be animated, the function can use countup; if not, it must insert the text with `insertText` just like it does for the `title`.

```
var boxxyBinding = new Shiny.OutputBinding();

$.extend(boxxyBinding, {
  find: function(scope) {
    return $(scope).find(".boxxy");
  },
  renderValue: function(el, data) {

    let boxValue, boxTitle;

    el.style.backgroundColor = data.color;

    if(data.animate){
      var counter = new CountUp(
        el.id + '-boxxy-value', 0, data.value
      );
      counter.start();
    } else {
      boxValue = document.getElementById(el.id + '-boxxy-value')
      boxValue.innerText = data.value;
    }

    boxTitle = document.getElementById(el.id + '-boxxy-title')
    boxTitle.innerText = data.title;
  }
});

Shiny.outputBindings.register(boxxyBinding, "john.boxxy");
```

Finally, we can render the dependency. The JavaScript method aptly named `renderDependencies` will do just that from the list of dependency created in `renderBoxxy`.

```
var boxxyBinding = new Shiny.OutputBinding();

$.extend(boxxyBinding, {
  find: function(scope) {
```

```
    return $(scope).find(".boxxy");
  },
  renderValue: function(el, data) {

    let boxValue, boxTitle;

    el.style.backgroundColor = data.color;

    if(data.animate){
      Shiny.renderDependencies(data.deps); // render dependency
      var counter = new CountUp(
        el.id + '-boxxy-value', 0, data.value
      );
      counter.start();
    } else {
      boxValue = document.getElementById(el.id + '-boxxy-value')
      boxValue.innerText = data.value;
    }

    boxTitle = document.getElementById(el.id + '-boxxy-title')
    boxTitle.innerText = data.title;
  }
});

Shiny.outputBindings.register(boxxyBinding, "john.boxxy");
```

With those changes made, not only is the animation of numbers optional, but
if users decide to turn off the animation in all boxxy functions, the countup.js
file will not be included at all.

```
library(shiny)

ui <- fluidPage(
  h2("Custom outputs"),
  fluidRow(
    column(
      3, boxxyOutput("countries")
    ),
    column(
      3, boxxyOutput("employees")
    )
  )
```

```
)

server <- function(input, output){
  output$countries <- renderBoxxy({
    boxxy("Countries", 176, animate = FALSE)
  })

  output$employees <- renderBoxxy({
    boxxy("Thing", 67, animate = FALSE)
  })
}

shinyApp(ui, server)
```

14.9 Preprocessing Custom Outputs

One aspect that this example did not truly explore thus far is the idea that the function boxxy should preprocess the input more in order to be truly justified. Currently boxxy only wraps the arguments in a list. Therefore, the code below works too.

```
# works too
output$theId <- renderBoxxy({
  list(
    title = "The Title",
    value = 123,
    color = "blue",
    animate = TRUE
  )
})
```

One the things boxxy could do is preprocess the input for instance. Instead of accepting a vector of length one for the value argument, it takes the sum of it to allow vectors of any length and dynamically changes the color depending on value.

```r
boxxy <- function(title, value, color = NULL, animate = TRUE){

  # sum the vector
  value <- sum(value)

  # dynamic color
  if(is.null(color))
    if(value > 100)
      color <- "#ef476f"
    else
      color <- "#06d6a0"

  list(
    title = title,
    value = value,
    color = color,
    animate = animate
  )
}
```

This means boxxy could be used like so: `boxxy("Total", c(1,6,9))`.

15

Custom Inputs

Shiny comes with a variety of inputs ranging from buttons to text fields; these inputs send data from the client to the R server. Custom inputs are in fact, no different than from Shiny's out-of-the-box inputs, they work in the same way and are built on the same system.

To explain and demonstrate how to build such a custom input, we shall build a switch input, which is essentially a fancy-looking checkbox that can be toggled on and off.

Custom Shiny inputs very much resemble Shiny outputs though they consist of a single R function (e.g.: `selectInput`), which generates the HTML and attaches necessary dependencies. When run from the R console, such functions will reveal the HTML they generate.

```
shiny::textInput("theId", "The label")
```

```
<div class="form-group shiny-input-container">
  <label class="control-label" for="theId">The label</label>
  <input id="theId" type="text" class="form-control" value=""/>
</div>
```

The R function is paired with a JavaScript input binding akin to the output binding used in the previous chapter.

15.1 Setup

Let us set up the necessary files and project structure. Below an asset directory is created. In it, we place a JavaScript file where the binding will be coded, as well as a CSS file that will style the switch input, an `app.R` file is also created to hold the R code and application.

DOI: 10.1201/9781003134046-15

```
# create directory for application
dir.create("switch/assets", recursive = TRUE)

# create R, JS, and CSS files
file.create("app.R")
file.create("switch/assets/binding.js")
file.create("switch/assets/styles.css")
```

This should create the following directory structure.

```
.
├── app.R
└── assets
    ├── binding.js
    └── styles.css
```

15.2   Switch Input HTML and Style

We will use w3schools[1]' switch input template.

The HTML of the input looks like so.

```
<label class="switch">
  <input type="checkbox">
  <span class="slider"></span>
</label>
```

While the CSS which stylises the checkbox input into a switch is the following.

```
.switch {
  position: relative;
  display: inline-block;
  width: 60px;
  height: 34px;
}
```

---

[1] https://www.w3schools.com/howto/howto_css_switch.asp

```
.switch input {
  opacity: 0;
  width: 0;
  height: 0;
}

.slider {
  position: absolute;
  cursor: pointer;
  top: 0;
  left: 0;
  right: 0;
  bottom: 0;
  background-color: #ccc;
  -webkit-transition: .4s;
  transition: .4s;
}

.slider:before {
  position: absolute;
  content: "";
  height: 26px;
  width: 26px;
  left: 4px;
  bottom: 4px;
  background-color: white;
  -webkit-transition: .4s;
  transition: .4s;
}

input:checked + .slider {
  background-color: #0462a1;
}

input:focus + .slider {
  box-shadow: 0 0 1px #0462a1;
}

input:checked + .slider:before {
  -webkit-transform: translateX(26px);
  -ms-transform: translateX(26px);
  transform: translateX(26px);
}
```

The above CSS should be placed in the previously created `assets/styles.css`
file. Figure 15.1 displays an unstyled checkbox and another styled with the
above CSS.

FIGURE 15.1: Checkbox and styled switch input

15.3   Generate Input HTML

Let us start with the R function to be used in the Shiny UI. The `<input>` it
generates bears a `switchInput` class, which will be used to identify all switch
inputs from JavaScript. This was also done in the custom output. The function
accepts an `id` argument; this is also common across all inputs and outputs as
a unique identifier is required in order to retrieve the input JavaScript-side.

```r
# app.R
switchInput <- function(id) {

  tags$input(
    id = id,
    type = "checkbox",
    class = "switchInput"
  )

}
```

The function should also accept a customary label and a `checked` argument to
define the initial state of the switch.

```r
# app.R
switchInput <- function(id, label, checked = TRUE) {

  input <- tags$input(
    id = id,
    type = "checkbox",
```

```
      class = "switchInput"
  )

  if(checked)
    input <- htmltools::tagAppendAttributes(input, checked = NA)

  form <- tagList(
    p(label),
    tags$label(
      class = "switch",
      input,
      tags$span(class = "slider")
    )
  )

  return(form)
}
```

As for the custom outputs, the dependencies (CSS and JavaScript binding) can piggy back on the generated HTML.

```
# app.R
switchInput <- function(id, label, checked = TRUE) {

  input <- tags$input(
    id = id,
    type = "checkbox",
    class = "switchInput"
  )

  if(checked)
    input <- htmltools::tagAppendAttributes(input, checked = NA)

  form <- tagList(
    p(label),
    tags$label(
      class = "switch",
      input,
      tags$span(class = "slider")
    )
  )

  path <- normalizePath("./assets")
```

```
deps <- htmltools::htmlDependency(
  name = "switchInput",
  version = "1.0.0",
  src = c(file = path),
  script = "binding.js",
  stylesheet = "styles.css"
)

htmltools::attachDependencies(form, deps)
}
```

15.4   JavaScript Input Binding

The JavaScript input binding is instantiated from `Shiny.InputBinding`, this is
similar to output bindings which are instantiated from `Shiny.OutputBinding`.

```
var switchInput = new Shiny.InputBinding();
```

Then again the binding is "extended." This consists of adding several methods.

- `find` returns all the relevant inputs.
- `getId` returns the unique identifier of the input.
- `getValue` returns the value of the input to be sent to the server.
- `setValue` is used to set the value of the input.
- `receiveMessage` is used to receive messages from the server.
- `subscribe` tells Shiny when and how to send the updated input value to the
  server.
- `unsubscribe` removes event handlers and stops Shiny from sending updated
  values to the server.

15.4.1   Find Inputs

The `find` method looks for all the relevant HTML elements in the document
(`scope`) and returns them as an array. Many other methods we are about to
implement will accept `el` as an argument; this will ultimately be one of the
elements returned by `find`.

Generally, the `find` method is used in conjunction with a `class`; hence the `<input>` generated by `switchInput` bears a class of the same name `switchInput`.

```
var switchInput = new Shiny.InputBinding();

$.extend(switchInput, {
  // find inputs
  find: function(scope) {
    return $(scope).find(".switchInput");
  }
});
```

15.4.2 Get Input Id

The `getId` method is exactly what it appears to be; it returns the `id` of the element. It looks for that id as `data-input-id` attribute, and if that is not found returns the `id`; this can be observed in the source code[2] (below).

```
this.getId = function(el) {
  return el['data-input-id'] || el.id;
};
```

Since the default works, there is no need to use it for the switch input.

15.4.3 Get Input Value

That retrieves the value of the input; this is often the attribute of the same name (`value="something"`), which can be obtained with the jQuery `val()` method. The switch is an input of type `checkbox` and therefore uses the `checked` prop.

```
var switchInput = new Shiny.InputBinding();

$.extend(switchInput, {
  find: function(scope) {
    return $(scope).find(".switchInput");
  },
  // retrieve value
```

---
[2]https://github.com/rstudio/shiny/blob/master/srcjs/input_binding.js#L9

```
  getValue: function(el) {
    return $(el).prop("checked");
  }
});
```

The value of the `checked` prop is boolean, `true` if checked and `false` if unchecked.

 Ensure the `getValue` method actually `returns` the value.

15.4.4   Set Input Value

The `setValue` method sets the value of the input; hence it also accepts the `value` object: the actual value to which the input should be set. Then again, most inputs will likely use the `value` attribute, which can be set in jQuery with `val(newValue)`, but the checkbox uses the `checked` prop.

```
var switchInput = new Shiny.InputBinding();

$.extend(switchInput, {
  find: function(scope) {
    return $(scope).find(".switchInput");
  },
  getValue: function(el) {
    return $(el).prop("checked");
  },
  // check or uncheck the switch
  setValue: function(el, value) {
    $(el).prop("checked", value).change();
  }
});
```

Note the use of the `change` method, which ensures the event is fired. Otherwise the input is checked or unchecked, but the `change` event is not fired, and this will cause problems later on as we rely on this event.

15.4.5   Receive Input Messages

The `setValue` method previously defined is only beneficial when combined with `receiveMessage`; the latter handles messages sent to the input, and these are generally sent from the server via functions the likes of `updateSelectInput`. Internally it uses the `setValue` method to define the value of the input received from the server. Note that the `value` is, therefore, a serialised JSON input coming from the R server and can be of any complexity you desire. Below we use it such that it expects a simple boolean as the checkbox (switch) can be either on (`true`) or off (`false`).

```javascript
var switchInput = new Shiny.InputBinding();

$.extend(switchInput, {
  find: function(scope) {
    return $(scope).find(".switchInput");
  },
  getValue: function(el) {
    return $(el).prop("checked");
  },
  setValue: function(el, value) {
    $(el).prop("checked", value).change();
  },
  // handle messages from the server
  receiveMessage: function(el, value){
    this.setValue(el, value);
  }
});
```

15.4.6 Subscribe and Unsubscribe Inputs

Finally, a crucial method is `subscribe`. This is run when the input is registered (more on that later) and is used to determine when Shiny sends new values of the input back to the server. This method also accepts a `callback`, which is the same function that tells Shiny to update the value. This callback function accepts a single boolean value, which the source code[3] states is used to enable debouncing or throttling. This is covered in the next section on rate policy.

This method often consists of an event listener that observes changes on the input to send it to the server. In layman terms, when the switch input changes (on to off or vice versa) run the `callback` function, which sends the data to the server.

---

[3] https://github.com/rstudio/shiny/blob/master/srcjs/input_binding.js#L18

```
var switchInput = new Shiny.InputBinding();

$.extend(switchInput, {
  find: function(scope) {
    return $(scope).find(".switchInput");
  },
  getValue: function(el) {
    return $(el).prop("checked");
  },
  setValue: function(el, value) {
    $(el).prop("checked", value).change();
  },
  receiveMessage: function(el, value){
    this.setValue(el, value);
  },
  subscribe: function (el, callback) {
    $(el).on("change.switchInput", function(){
      callback(true);
    })
  },
  unsubscribe: function(el) {
    $(el).off(".switchInput");
  }
});
```

Note that in the `subscribe` method we listen for `changes` on the input; hence the `setValue` also uses jQuery's `change` method; it ensures this event is fired and that Shiny will subsequently pick it up.

Make sure the `setValue` method triggers the event observed in `subscribe`

15.4.7 Input Rate Policy

The rate policy determines how frequently the binding should send new input values back to the server. The `getRatePolicy` method should return an object that describes a JSON array with two variables: `policy` and `delay`.

direct

The `direct` policy tells Shiny to sends any new value directly, however often this occurs. Therefore, this policy does not make use of `delay`.

```
{
  policy: "direct"
}
```

debounce

The `debounce` policy tells Shiny to ignore all new values until no new values
have been received for `delay` milliseconds.

```
{
  policy: "debounce",
  delay: 2500
}
```

throttle

The `throttle` policy means that no more than one value will be sent per `delay`
milliseconds.

```
{
  policy: "throttle",
  delay: 1000
}
```

A switch input is not expected to change frequently, but it's nonetheless good
practice to throttle it to ensure the server does not receive too many requests.
This will admittedly be more relevant to inputs that see a higher rate of change
like text fields, which unless debounced send every keystroke to the server.

```
var switchInput = new Shiny.InputBinding();

$.extend(switchInput, {
  find: function(scope) {
    return $(scope).find(".switchInput");
  },
  getValue: function(el) {
    return $(el).prop("checked");
  },
  setValue: function(el, value) {
    $(el).prop("checked", value).change();
```

```
    },
    receiveMessage: function(el, value){
      this.setValue(el, value);
    },
    subscribe: function (el, callback) {
      $(el).on("change.switchInput", function(){
        callback(true);
      })
    },
    unsubscribe: function(el) {
      $(el).off(".switchInput");
    },
    // throttle
    getRatePolicy: function(){
      return {
        policy: 'throttle',
        delay: 1000
      }
    }
});
```

15.4.8   Registering the Input Binding

Finally, like the custom output, the input can be registered with Shiny. It too
takes a unique identifier as a second argument.

```
var switchInput = new Shiny.InputBinding();

$.extend(switchInput, {
  find: function(scope) {
    return $(scope).find(".switchInput");
  },
  getValue: function(el) {
    return $(el).prop("checked");
  },
  setValue: function(el, value) {
    $(el).prop("checked", value).change();
  },
  receiveMessage: function(el, value){
    this.setValue(el, value);
  },
```

```
  subscribe: function (el, callback) {
    $(el).on("change.switchInput", function(){
      callback(true);
    })
  },
  unsubscribe: function(el) {
    $(el).off(".switchInput");
  },
  getRatePolicy: function(){
    return {
      policy: 'throttle',
      delay: 1000
    }
  }
});

Shiny.inputBindings.register(switchInput, 'john.switch');
```

This wraps up a custom input: it can now be used in a shiny application (see Figure 15.2).

```
library(shiny)

ui <- fluidPage(
  switchInput("switch", "Show plot", FALSE),
  plotOutput("plot")
)

server <- function(input, output, session){

  output$plot <- renderPlot({
    print(input$switch)

    if(!input$switch)
      return()

    plot(cars)
  })
}

shinyApp(ui, server)
```

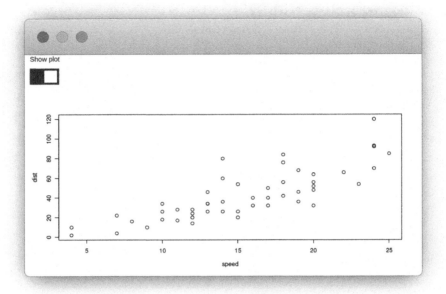

FIGURE 15.2: Switch input example

15.5   Update Input

The JavaScript binding was provided with the required methods to receive updates from the server to change the value of the switch input, but the R function that sends those updates is yet to be written.

```r
update_switch_input <- function(id, value,
  session = shiny::getDefaultReactiveDomain()){
  session$sendInputMessage(id, value)
}
```

This pattern was used previously, albeit using `sendCustomMessage`; with `sendInputMessage` the message can be sent straight to the `receiveMessage` handler of the input using 1) the `id` of the input and 2) the data one wants to send.

We can adapt the application to use this button.

```r
library(shiny)

ui <- fluidPage(
  actionButton("chg", "Switch ON"),
  switchInput("switch", "Switch input", FALSE),
  plotOutput("plot")
)

server <- function(input, output, session){

  output$plot <- renderPlot({
    print(input$switch)

    if(!input$switch)
      return()

    plot(cars)
  })

  observeEvent(input$chg, {
    update_switch_input("switch", TRUE, session)
  })
}

shinyApp(ui, server)
```

Figure 15.3 attempts to summarize the various elements that were put together and used in the last application.

It all starts from the `switchInput` function, which generates the HTML defining the switch input and its initial state. In the `subscribe` method, an event listener checks for changes on this HTML element (`$(el).on('change', ...)`). Every time it changes (check/uncheck) it fires the Shiny `callback`, which sends the value of the input obtained from `getValue` through the WebSocket. When the value of the input is changed from the server this value travels through the WebSocket to the front end, where `receiveMessage` uses `setValue` to programmatically change the check-box, which incidentally triggers the change event, and back we go.

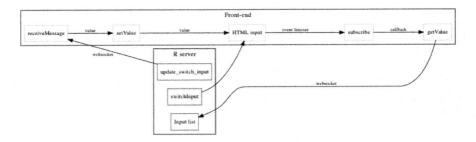

FIGURE 15.3: Shiny input visualised

If you wish to explore Shiny inputs in more depth I suggests reading David Granjon's book: *Outstanding Shiny UI.*[4]

15.6 Exercise

- Add a `toggle_switch_input` function that toggles between states so it turns it off when it's on and vice versa
- Bring support for Bootstrap 3 text input addon[5]

---

[4]https://unleash-shiny.rinterface.com/

[5]https://getbootstrap.com/docs/3.3/components/#input-groups-basic

16

Cookies

In this chapter, we scout yet another excellent example of how JavaScript can enhance Shiny applications. We use an HTTP cookie, a small piece of data sent from an application and stored in the user's web browser, to track users returning to a Shiny application.

The application will prompt users to input their name; this will be stored in a cookie so that on their next visit, they are welcomed to the app with a personalised message. Cookies are natively supported by web browsers and JavaScript, though we will use a library which greatly eases their handling: js-cookie[1].

16.1 Discover js-cookie

The library is at its core very straightforward; it exports a `Cookie` object from which one can access the `set`, `get`, and `remove` methods.

```
// set a cookie
Cookies.set('name', 'value')

// get cookies
Cookies.get();

// remove a cookie
Cookies.remove('name');
```

There is also the possibility to pass additional options when defining the cookie, such as how long it is valid for, but we won't explore these here.

---

[1] https://github.com/js-cookie/js-cookie

DOI: 10.1201/9781003134046-16

16.2   Setup Project

Then again, it starts with the creation of a directory where we'll place a
JavaScript file containing the message handlers; we won't download the dependency and use the CDN instead but feel free to do differently.

```r
dir.create("www")
file.create("www/script.js")
```

We then lay down the skeleton of the application which features a text input
to capture the name of the user, a button to save the cookie in their browsers,
another to remove it and finally a dynamic output that will display the
personalised message.

```r
library(shiny)

addResourcePath("www", "www")

ui <- fluidPage(
  tags$head(
    tags$script(
      src = paste0(
        "https://cdn.jsdelivr.net/npm/js-cookie@rc/",
        "dist/js.cookie.min.js"
      )
    ),
    tags$script(src = "www/script.js")
  ),
  textInput("name_set", "What is your name?"),
  actionButton("save", "Save cookie"),
  actionButton("remove", "remove cookie"),
  uiOutput("name_get")
)

server <- function(input, output, session){

}

shinyApp(ui, server)
```

16.3   JavaScript Cookies

First, we define a JavaScript function that retrieves the cookies and sets the result as an input named `cookies`. The reason we do so is because we will have to execute this in multiple places; this will become clearer in just a second.

```
// script.js
function getCookies(){
  var res = Cookies.get();
  Shiny.setInputValue('cookies', res);
}
```

Then we define two message handlers, one that sets the cookie and another that removes it. Note that both of them run the `getCookies` function defined previously after they are done with their respective operations. The reason this is done is that we need the input to be updated with the new values after it is set and after it is removed. Otherwise setting or removing the cookie will leave the actual input value untouched and the result of the operation (setting or removing the cookie) will not be captured server-side.

```
// script.js
Shiny.addCustomMessageHandler('cookie-set', function(msg){
  Cookies.set(msg.name, msg.value);
  getCookies();
})

Shiny.addCustomMessageHandler('cookie-remove', function(msg){
  Cookies.remove(msg.name);
  getCookies();
})
```

One more thing needs to be implemented JavaScript-side. The point of using the cookie is that when users come back to the Shiny app, even days later, they are presented with the personalised welcome message, this implies that when they open the application, the input value is defined. Therefore, `getCookies` must at launch. One could be tempted to place it at the top of the JavaScript file so that it runs when it is imported but the issue there is that it will run too soon, it will run the Shiny input is not yet available and fail to set it. Consequently, we instead observe the `shiny:connected` event, which is fired when the initial connection to the server is established.

```
// script.js
$(document).on('shiny:connected', function(ev){
  getCookies();
})
```

16.4   R Code

Then it's a matter of completing the Shiny server which was left empty. We add
an observeEvent on the save button where we check that a name has actually
been typed in the text box before saving the cookie. There is another similar
observer on the remove button. The renderUI expression checks that the cookie
has been set and displays a message accordingly.

```r
library(shiny)

addResourcePath("www", "www")

ui <- fluidPage(
  tags$head(
    tags$script(
      src = paste0(
        "https://cdn.jsdelivr.net/npm/js-cookie@rc/",
        "dist/js.cookie.min.js"
      )
    ),
    tags$script(src = "www/script.js")
  ),
  textInput("name_set", "What is your name?"),
  actionButton("save", "Save cookie"),
  actionButton("remove", "remove cookie"),
  uiOutput("name_get")
)

server <- function(input, output, session){

  # save
  observeEvent(input$save, {
    msg <- list(
      name = "name", value = input$name_set
```

```
   )

   if(input$name_set != "")
     session$sendCustomMessage("cookie-set", msg)
})

# delete
observeEvent(input$remove, {
  msg <- list(name = "name")
  session$sendCustomMessage("cookie-remove", msg)
})

# output if cookie is specified
output$name_get <- renderUI({
  if(!is.null(input$cookies$name))
    h3("Hello,", input$cookies$name)
  else
    h3("Who are you?")
})

}

shinyApp(ui, server)
```

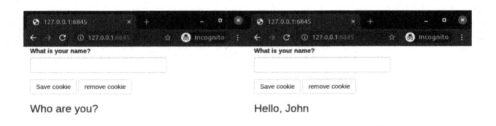

FIGURE 16.1: Shiny using cookies

Run the application and save your name. You can then refresh the application, and the welcome message will display your name. You can even kill the server entirely and re-run the app; the welcome message will still display as in Figure 16.1!

16.5   Exercises

In the following chapter, we cover the use of Shiny and bidirectional specifically
for htmlwidgets. Before moving on it is a good idea to attempt to replicate the
examples that were created in this part of the book. Below are some interesting
JavaScript libraries that are not yet integrated with Shiny (as packages). These
would be great additions to the Shiny ecosystem and are specifically selected
because they are approachable.

- micromodal.js[2] - tiny, dependency-free javascript library for creating accessible modals
- hotkeys[3] - capture keyboard inputs
- handtrack.js[4] - realtime hand detection
- annyang[5] - speech Recognition library

---

[2] https://github.com/Ghosh/micromodal

[3] https://github.com/jaywcjlove/hotkeys

[4] https://github.com/victordibia/handtrack.js

[5] https://github.com/TalAter/annyang

17

Widgets with Shiny

We have seen how to make JavaScript and R communicate in Shiny applications by passing data from the server to the client and back. This chapter explores how to apply that to htmlwidgets so they can provide additional functionalities when used in Shiny applications.

To demonstrate how to integrate these functionalities in widgets, we shall implement them in the previously built gio package.

17.1 Widgets to R

In a previous chapter on Linking Widgets, the topic of callback function was briefly explored. Gio.js provides a callback function that is fired every time a country is selected on globe; this is currently used to share said country with other widgets as part of the implementation with crosstalk. This callback function could also send data back to the R server where they could be used for many things like fetching more data on the selected country from a database or use that information to generate a Shiny UI element like displaying the flag of the selected country, and much, much more.

The callback function in the gio package currently looks like this; it only makes use of some of the data for crosstalk.

```
// define callback function
function callback (selectedCountry, relatedCountries) {
  sel_handle.set([selectedCountry.ISOCode]);
}

// use callback function
controller.onCountryPicked(callback);
```

We can make further use of this to define two different Shiny inputs; one for the selected country and another for its related edges.

```
// gio.js
// define callback function
function callback (selectedCountry, relatedCountries) {
  sel_handle.set([selectedCountry.ISOCode]);
  Shiny.setInputValue('selected', selectedCountry);
  Shiny.setInputValue('related', relatedCountries);
}
```

However, this will generate an issue experienced in a previous chapter; multiple
gio visualisations in a single Shiny application would be defining the values of
a single input. This can be remedied to by using the id of the visualisation to
generate the input name dynamically.

```
renderValue: function(x) {

  var controller = new GIO.Controller(el);
  controller.addData(x.data);
  controller.setStyle(x.style);

  // callback
  function callback (selectedCountry, relatedCountries) {
    sel_handle.set([selectedCountry.ISOCode]);
    Shiny.setInputValue(el.id + '_selected', selectedCountry);
    Shiny.setInputValue(el.id + '_related', relatedCountries);
  }

  controller.onCountryPicked(callback);

  // render
  controller.init();

}
```

The package can then be installed with devtools::install so we can test these
inputs in a Shiny application (Figure 17.1).

```
library(gio)
library(shiny)

# large sample data
url <- paste0(
```

```
   "https://raw.githubusercontent.com/JohnCoene/",
   "javascript-for-r/master/data/countries.json"
)
arcs <- jsonlite::fromJSON(url)

ui <- fluidPage(
  gioOutput("globe"),
  fluidRow(
    column(6, verbatimTextOutput("selectedCountry")),
    column(6, verbatimTextOutput("relatedCountries"))
  )
)

server <- function(input, output){

  output$globe <- renderGio({
    gio(arcs, source = i, target = e, value = v)
  })

  output$selectedCountry <- renderPrint({
    print(input$globe_selected)
  })

  output$relatedCountries <- renderPrint({
    print(input$globe_related)
  })

}

shinyApp(ui, server)
```

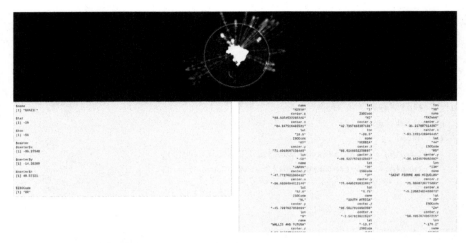

FIGURE 17.1: Gio with input data

One thing to note before moving on, the data is sent from the client to the server whether the inputs are used or not, though this likely will not negatively impact gio it can reduce performances if the callback function is triggered too frequently. For instance, an input value set when the user hovers a scatter plot might lead to the event being fired very frequently and too much data being sent to the server, slowing things down and providing a poor experience.

Therefore one might consider making the capture of such event optional, so the web browser is not strained unless explicitly asked by the developer of the application. This could be implemented with a simple function that sets a simple logical variable in the x object that is used in JavaScript to check whether to implement the callback function.

```
#' @export
gio_capture_events <- function(g) {
  g$x$capture_events <- TRUE
  return(g)
}
```

Then this could be used in JavaScript with an if statement.

```
if(x.capture_events)
  controller.onCountryPicked(callback);
```

One might also consider not sending back all the data. For instance, gio returns the coordinates of the selected country wherefrom arcs connect; this might be considered unnecessary. The code below only sets the input to the ISO code of the country selected.

```
function callback (selectedCountry, relatedCountries) {
  Shiny.setInputValue(
    el.id + '_selected',
    selectedCountry.ISOCode
  );
}
```

17.2   Input Handlers for Widgets

Input handlers were explored in an earlier chapter where it was used to transform the results of an image classification model from a list to a data.frame. We can apply something very similar to this input data; the "related countries" returned consist of edges and should be reshaped into a data.frame as well (that looks like gio's input data).

Below we create a handler that is going to loop over the list (over each arc) and return a data frame.

```
# handler.R
related_countries_handler <- function(x, session, inputname){
  purrr::map_dfr(x, as.data.frame)
}
```

Then the handler must be registered with Shiny since handlers can only be registered once an excellent place to put it this is in the .onLoad function of the package.

```
# zzz.R
related_countries_handler <- function(x, session, inputname){
  purrr::map_dfr(x, as.data.frame)
}

.onLoad <- function(libname, pkgname) {
```

```
  shiny::registerInputHandler(
    "gio.related.countries",
    related_countries_handler
  )
}
```

Finally, we can reinstall the package with `devtools::install` and create a Shiny application to observe the change. In Figure 17.2 we use a large example dataset and, since the input now returns a data frame, we can display the input value in a table.

```
library(DT)
library(gio)
library(shiny)

# large sample data
url <- paste0(
  "https://raw.githubusercontent.com/JohnCoene/",
  "javascript-for-r/master/data/countries.json"
)
arcs <- jsonlite::fromJSON(url)

ui <- fluidPage(
  gioOutput("globe"),
  DTOutput("relatedCountries")
)

server <- function(input, output){

  output$globe <- renderGio({
    gio(arcs, i, e, v)
  })

  output$relatedCountries <- renderDT({
    datatable(input$globe_related)
  })

}

shinyApp(ui, server)
```

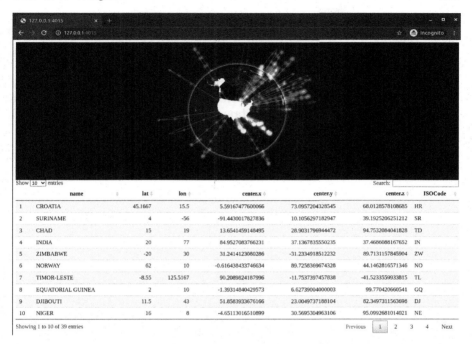

	name	lat	lon	center.x	center.y	center.z	ISOCode
1	CROATIA	45.1667	15.5	5.59167477600066	73.0957204328545	68.0128578108685	HR
2	SURINAME	4	-56	-91.4430017827836	10.1056297182947	39.1925206251212	SR
3	CHAD	15	19	13.6541459148495	28.9031796944472	94.7532084041828	TD
4	INDIA	20	77	84.9527083766231	37.1367835550235	37.4686088167652	IN
5	ZIMBABWE	-20	30	31.2414123080286	-31.2334918512232	89.7131157845904	ZW
6	NORWAY	62	10	-0.616438433746634	89.7258369674328	44.1462816571346	NO
7	TIMOR-LESTE	-8.55	125.5167	90.2089824187996	-11.7537397457838	-41.5233559933815	TL
8	EQUATORIAL GUINEA	2	10	-1.39314840429573	6.62739004000003	99.770420660541	GQ
9	DJIBOUTI	11.5	43	51.8583933676166	23.0049737188104	82.3497311563698	DJ
10	NIGER	16	8	-4.65113016510899	30.5695304963106	95.0992681014021	NE

Showing 1 to 10 of 39 entries   Previous 1 2 3 4 Next

FIGURE 17.2: Gio input data transformed to a data frame

17.3 R to Widgets

This book previously explored how to send data from the Shiny server to the front end; this section applies this to htmlwidgets. Currently, using gio in Shiny consists of generating the globe with the `renderGio` and complimentary `gioOutput` functions. This generates the complete visualisation, it creates the HTML element where it places the globe, draws the arcs based on the data, sets the style, etc.

Now imagine that only one of those aspects needs changing, say the data, or the style, given the functions currently at hand one would have to redraw the entire visualisation, only this time changing the data or the style. This is inelegant and not efficient, most JavaScript visualisation libraries, including gio.js, will enable changing only certain aspects of the output without having to redraw it all from scratch.

Before we look into the implementation, let us create a Shiny application which would benefit from such a feature. The Shiny application below (Figure 17.3)

provides a drop-down menu to select between two datasets to draw on the globe, running it reveals an issue with gio though. Upon selecting a dataset with the drop down a second globe appears underneath the original one. This is because internally gio.js creates a new element (`<canvas>`) within the `<div>` created by `htmlwidgets` when running `init` regardless of whether one was already created. Therefore, every call to `init` creates a new `<canvas>` with a different globe. Note that most visualisation libraries *will not have that issue*, they will detect the existing output and override it instead.

```r
library(gio)
library(shiny)

arcs1 <- data.frame(
  e = c("US", "CN", "RU"),
  i = c("CN", "RU", "US"),
  v = c(100, 120, 130)
)

arcs2 <- data.frame(
  e = c("CN", "CN", "JP"),
  i = c("IN", "JP", "US"),
  v = c(100, 120, 130)
)

ui <- fluidPage(
  selectInput(
    "dataset",
    "Select a dataset",
    choices = c("First", "Second")
  ),
  gioOutput("globe")
)

server <- function(input, output){

  reactive_arcs <- reactive({
    if(input$dataset == "First")
      return(arcs1)
    return(arcs2)
  })

  output$globe <- renderGio({
    gio(reactive_arcs(), i, e, v)
  })
```

```
}

shinyApp(ui, server)
```

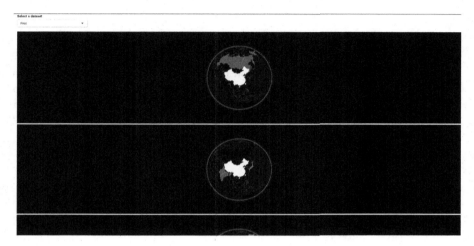

FIGURE 17.3: Gio issue in shiny

A solution to this is to ensure the container (el) is empty before generating the visualisation. Incidentally, this can be executed with a JavaScript method previously used in this book: innerHTML.

```
// gio.js
el.innerHTML = ''; // empty el
controller = new GIO.Controller(el);
```

Now, using the dropdown to switch between dataset does not generate a new visualisation.

We got sidetracked, but this had to be fixed. Ideally, when the user selects a dataset from the dropdown the entire visualisation is not redrawn, only the underlying data (the arcs) changes. To do so, a new set of functions divorced from the ones currently at hand needs to be created. This separation will allow leaving the already created functions as-is, using gio and its corresponding renderValue JavaScript function to initialise a visualisation, and have a separate family of functions dedicated to working with different JavaScript functions which circumvent renderValue and directly change aspects of the visualisation, such as the underlying dataset.

This involves a few moving parts, thankfully some of them were already explored, just not in the context of htmlwidgets. The scheme is to send data from R to JavaScript using the formerly exploited session$sendCustomMessage, then in JavaScript fetch the instance of the visualisation (controller in the case of gio) to interact with it (e.g. controller.addData(data);).

17.3.1 Send Data from Widgets

Let us start by creating the R function used to send data to JavaScript. This is hopefully reminiscent of a previous chapter; this function sends a message to JavaScript where it is paired with "message handler" that does something with said message. Note that for consistency the function uses the same arguments as the other function of the package which accepts data: gio.

```
#' @export
gio_send_data <- function(id, data, source, target, value,
  session = shiny::getDefaultReactiveDomain()){

  data <- dplyr::select(
    data,
    i = {{ source }},
    e = {{ target }},
    v = {{ value }}
  )
  message <- list(id = id, data = data)
  session$sendCustomMessage("send-data", message)
}
```

The function takes the id of the visualisation the data is destined for, the data object itself, and a Shiny session used to send the data. The id of the visualisation is sent as part of the message and will be used to retrieve the instance of the visualisation and subsequently apply the new dataset. Note that we give this message the send-data identifier, this will be needed when we write its handler.

There is one caveat that will make it such that the above will not work. To have gio.js work with data changes had to be made to the serialiser (using the TOJSON_ARGS attribute), this cannot be used here. The data being sent with Shiny via the session object the problem reoccurs: Shiny, like htmlwidgets, serialises data frames column-wise and not row-wise. One can preview the way Shiny serialises with shiny:::toJSON (three-colon).

```
# preview shiny serialisation
shiny:::toJSON(arcs)
#> {"e":["CN","CN"],"i":["US","RU"],"v":[3300000,10000]}
```

Unfortunately this serialiser cannot be changed, therefore we have to reformat the data to a list which resembles the JSON output desired, using `apply` to turn every row into a list will do the job in most cases.

```
#' @export
gio_send_data <- function(id, data, source, target, value,
  session = shiny::getDefaultReactiveDomain()){

  data <- dplyr::select(
    data,
    i = {{ source }},
    e = {{ target }},
    v = {{ value }}
  )
  message <- list(id = id, data = apply(data, 1, as.list))
  session$sendCustomMessage("send-data", message)
}
```

17.3.2 Retrieve Widget Instance

We will need to be able to access the instance of the visualisation (`controller`) outside of the function `factory`. This can be made accessible by adding a function (technically a method) that returns the `controller`. Below we create a function called `getGlobe` which returns the `controller`.

```
HTMLWidgets.widget({

  name: 'gio',

  type: 'output',

  factory: function(el, width, height) {

    // TODO: define shared variables for this instance
    var controller;
```

```
// selection handle
var sel_handle = new crosstalk.SelectionHandle();

sel_handle.on("change", function(e) {
  if (e.sender !== sel_handle) {
    // clear selection
  }
  controller.switchCountry(e.value[0]);
});


return {

  renderValue: function(x) {

    el.innerHTML = '';
    controller = new GIO.Controller(el, x.configs);

    // group
    sel_handle.setGroup(x.crosstalk.group);

    // add data
    controller.addData(x.data);

    controller.setStyle(x.style);

    // callback
    controller.onCountryPicked(callback);

    function callback (selectedCountry, relatedCountries) {
      sel_handle.set([selectedCountry.ISOCode]);
      Shiny.setInputValue(
        el.id + '_selected',
        selectedCountry
      );
      Shiny.setInputValue(
        el.id + '_related:gio.related.countries',
        relatedCountries
      );
    }

    // use stats
    if(x.stats)
```

```
            controller.enableStats();

        // render
        controller.init();

    },

    resize: function(width, height) {

        // TODO: code to re-render the widget with a new size
        controller.resizeUpdate()

    },

    getGlobe: function(){
        return controller;
    }

    };
  }
});
```

Then, anywhere in the gio.js file, we create a function that uses HTMLWidgets.find to retrieve the instance of the htmlwidgets it takes a selector as input. We concatenate the pound sign to select the widget based on its id (#id, .class). This, in effect, returns an object, which includes all the functions returned by the factory function: renderValue, resize, and getGlobe. We can therefore use the getGlobe method available from that object to retrieve the actual controller.

```
// gio.js
function get_gio(id){
    var widget = HTMLWidgets.find("#" + id);
    var globe = widget.getGlobe();
    return globe;
}
```

17.3.3 Handle Data

We can now turn our attention to actually applying the data sent from the R server to the visualisation: the "message handler." Registering the message handler is only relevant if Shiny is running. Therefore htmlwidgets comes with a function to check whether that is the case, which is useful to avoid needless errors. We can thus use it in an if-statement in which all message handlers will be registered.

```
// gio.js

// check if shiny running
if (HTMLWidgets.shinyMode){

  // send-data message handler
  Shiny.addCustomMessageHandler(
    type = 'send-data', function(message) {

  });

}
```

What the currently empty send-data message handler should do is fetch the widget using the id sent from R with the get_gio function and then use the addData method to override the previously-defined arcs.

```
// gio.js

// check if shiny running
if (HTMLWidgets.shinyMode){

  // send-data message handler
  Shiny.addCustomMessageHandler(
    type = 'send-data', function(message) {

    // retrieve controller
    var controller = get_gio(message.id);

    // add data
    controller.addData(message.data);

  });

}
```

We can then build a Shiny application to test the new functionality.

```r
library(gio)
library(shiny)

# two phoney datasets
arcs1 <- data.frame(
  e = c("US", "CN", "RU"),
  i = c("CN", "RU", "US"),
  v = c(100, 120, 130)
)

arcs2 <- data.frame(
  e = c("CN", "CN", "JP"),
  i = c("IN", "JP", "US"),
  v = c(100, 120, 130)
)

ui <- fluidPage(
  selectInput(
    "dataset",
    "Select a dataset",
    choices = c("First", "Second")
  ),
  gioOutput("globe")
)

server <- function(input, output){

  reactive_arcs <- reactive({
    if(input$dataset == "First")
      return(arcs1)
    return(arcs2)
  })

  output$globe <- renderGio({
    gio(arcs1, i, e, v)
  })

  observeEvent(input$dataset, {
    if(input$dataset == "First")
      data <- arcs1
    else
      data <- arcs2

    gio_send_data("globe", data, i, e, v)
```

```
  })

}

shinyApp(ui, server)
```

Switching dataset with the dropdown only changes the data; it makes for a much smoother animation, even the difference in the speed at which the effect is visible on the visualisation is perceivable.

17.4 Proxy Function

Before we add other similar functions, we ought to pause and consider the API this provides the user. There are two points, every function such as gio_send_data, will need to accept the id and session arguments. It will be tedious to so every time, following the old "don't repeat yourself" adage we ought to abstract this further.

This can be remedied to by introducing what is often referred to as a "proxy." A proxy is just a representation of the graph, or pragmatically, an object that contains the id of the visualisation and a Shiny session. This object can subsequently be piped to other functions, thereby providing not only a cleaner but also a more consistent API.

```
#' @export
gio_proxy <- function(id,
  session = shiny::getDefaultReactiveDomain()
){

  list(id = id, session = session)
}
```

Above we create a function called gio_proxy; it takes the id of the chart one wants to build a proxy for, as well as the Shiny session, these are returned in the form of a list. Next, we should adapt the gio_send_data so that it accepts the output of gioProxy instead of the id and session as done previously. In order to allow chaining such functions, we also make sure it returns the proxy object.

```
#' @export
gio_send_data <- function(proxy, data){
  message <- list(id = proxy$id, data = apply(data, 1, as.list))
  proxy$session$sendCustomMessage("send-data", message)
  return(proxy)
}
```

17.5   Clear Data

In order to actively demonstrate the advantage of the "proxy" function
as well as to hammer home how such functions work, we shall build another, which removes all data from the globe. In JavaScript, it's as simple as
`controller.clearData();`.

The journey starts with the R code, where we create a new function that sends
a message to clear the data; the message only needs to contain the id of the
visualisation from which data needs to be cleared, as before it will be used to
retrieve the `controller` from which the `clearData` method is available.

```
#' @export
gio_clear_data <- function(proxy){
  message <- list(id = proxy$id)
  proxy$session$sendCustomMessage("clear-data", message)
  return(proxy)
}
```

Now onto the JavaScript code to catch that message and actually clear the
data from the globe. That function is very similar to what was previously
shown, the only difference is the name of the message handler and the method
used on the controller.

```
// gio.js
Shiny.addCustomMessageHandler(
  type = 'clear-data', function(message) {

    var controller = get_gio(message.id);
```

```
      controller.clearData();
});
```

Then one can build an application to test that new function. We build a Shiny application with a button to add the data to the visualisation and another to clear data from it, as shown in Figure 17.4.

```r
library(gio)
library(shiny)

# phoney dataset
arcs <- data.frame(
  e = c("US", "CN", "RU"),
  i = c("CN", "RU", "US"),
  v = c(100, 120, 130)
)

ui <- fluidPage(
  actionButton("load", "Load data"),
  actionButton("clear", "Clear data"),
  gioOutput("globe")
)

server <- function(input, output){

  output$globe <- renderGio({
    gio(arcs, i, e, v)
  })

  observeEvent(input$load, {
    gio_proxy("globe") %>%
      gio_send_data(arcs, i, e, v)
  })

  observeEvent(input$clear, {
    gio_proxy("globe") %>%
      gio_clear_data()
  })

}

shinyApp(ui, server)
```

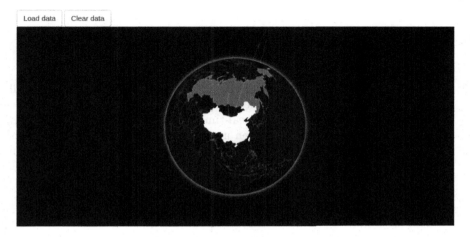

FIGURE 17.4: Gio with clear data proxy

17.6 Update the Widget

The previous proxies defined worked for reasons unbeknownst to the author of this book. It will not work with all methods. The reason it will not work is one that is likely to occur with many other visualisation libraries. For instance, one can attempt to develop a function to dynamically change the style without having to redraw the entire globe, starting again with the R function.

```r
#' @export
gio_set_style <- function(proxy, style){
  message <- list(id = proxy$id, style = style)
  proxy$session$sendCustomMessage("set-style", message)
  return(proxy)
}
```

Then, adding the JavaScript handler.

```js
Shiny.addCustomMessageHandler(
  type = 'set-style', function(message) {
    var controller = get_gio(message.id);
```

```
    controller.setStyle(message.style);
});
```

At this stage, one can try the function in a Shiny application, but it will not work because most such methods that change underlying aspects of a visualisation will not be reflected in real-time. Gio.js, and many other libraries, will require one to explicitly ask for an update so the changes take effect. This has multiple advantages, one can stack multiple visual changes to execute them at the same time, and one can manage the load on the front end.

```
Shiny.addCustomMessageHandler(
  type = 'set-style', function(message) {
    var controller = get_gio(message.id);
    controller.setStyle(message.style);
    controller.update(); // force update the visualisation
});
```

This forces the chart to update, applying the new style. Below we write an application that provides a dropdown to switch between two styles (Figure 17.5).

```
library(gio)
library(shiny)

# two phoney datasets
arcs <- data.frame(
  e = c("US", "CN", "RU"),
  i = c("CN", "RU", "US"),
  v = c(100, 120, 130)
)

ui <- fluidPage(
  selectInput(
    "style",
    "Select a style",
    choices = c("blueInk", "earlySpring")
  ),
  gioOutput("globe")
)

server <- function(input, output){
```

```
  output$globe <- renderGio({
    gio(arcs1, i, e , v)
  })

  observeEvent(input$style, {
    gio_proxy("globe") %>%
      gio_set_style(input$style)
  })

}
shinyApp(ui, server)
```

Select a style

blueink

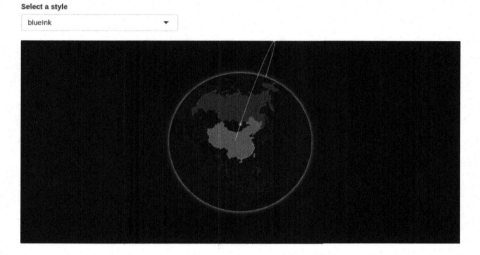

FIGURE 17.5: Gio with dynamic style

Part IV

JavaScript for Computations

18

The V8 Engine

V8 is an R interface to Google's open-source JavaScript engine of the same name; it powers Google Chrome, Node.js, and many other things. It is the last integration of JavaScript with R that is covered in this book. Both the V8 package and the engine it wraps are straightforward yet amazingly powerful.

18.1 Installation

First, install the V8 engine itself, instructions to do so are well detailed on V8's README[1] and below.

On Debian or Ubuntu use the code below from the terminal to install libv8[2].

```
sudo apt-get install -y libv8-dev
```

On Centos install v8-devel, which requires the EPEL tools.

```
sudo yum install epel-release
sudo yum install v8-devel
```

On Mac OS use Homebrew[3].

```
brew install v8
```

Then install the R package from .

---

[1] https://github.com/jeroen/v8#installation
[2] https://v8.dev/
[3] https://brew.sh/

DOI: 10.1201/9781003134046-18   255

```
install.packages("V8")
```

18.2   Basics

V8 provides an JavaScript execution environment through returning a closure-
based object with v8(); Each of such environments is independent of another.

```
library(V8)

engine <- v8()
```

The eval method allows running JavaScript code from R.

```
engine$eval("var x = 3 + 4;") # this is evaluated in R
engine$eval("x")
#> [1] "7"
```

Two observations are worth making on the above snippet of code. First, the
variable we got back in R is a character vector when it should have been either
an integer or a numeric. This is because we used the eval method, which returns
what is printed in the V8 console, but get is more appropriate; it converts the
output to its appropriate R equivalent.

```
# retrieve the previously created variable
(x <- engine$get("x"))
#> [1] 7
class(x)
#> [1] "integer"
```

Second, while creating a scalar with eval("var x = 1;") appears painless, imagine
if you will the horror of having to convert a data frame to a JavaScript array
via jsonlite then flatten it to character string so it can be used with the eval
method. Horrid. Thankfully V8 comes with a method assign, complimentary to
get, which declares R objects as JavaScript variables. It takes two arguments,
first the name of the variable to create, second the object to assign to it.

```
# assign and retrieve a data.frame
engine$assign("vehicles", cars[1:3, ])
engine$get("vehicles")
#>    speed dist
#> 1     4    2
#> 2     4   10
#> 3     7    4
```

All of the conversion is handled by V8 internally with jsonlite, as demonstrated in the previous chapter. We can confirm that the data frame was converted to a list row-wise, using JSON.stringify to display how the object is stored in V8.

```
cat(engine$eval("JSON.stringify(vehicles, null, 2);"))
#> [
#>   {
#>     "speed": 4,
#>     "dist": 2
#>   },
#>   {
#>     "speed": 4,
#>     "dist": 10
#>   },
#>   {
#>     "speed": 7,
#>     "dist": 4
#>   }
#> ]
```

However this reveals a tedious cyclical loop: 1) creating an object in JavaScript to 2) run a function on the aforementioned object, 3) get the results back in R, and repeat. So V8 also allows calling JavaScript functions on R objects directly with the call method and obtains the results back in R.

```
engine$eval("new Date();") # using eval
```

```
#> [1] "Sun Oct 18 2020 18:34:45 GMT+0200
#>    (Central European Summer Time)"
```

```
engine$call("Date", Sys.Date()) # using call
```

```
#> [1] "Sun Oct 18 2020 18:34:45 GMT+0200
   (Central European Summer Time)"
```

Finally, one can run code interactively rather than as strings by calling the console from the engine with `engine$console()`. You can then exit the console by typing `exit` or hitting the ESC key.

18.3   External Libraries

V8 is quite bare in and of itself; there is, for instance, no functionalities built in to read or write files from disk. It thus becomes truly interesting when you can leverage JavaScript libraries. We'll demonstrate this using fuse.js[4] a fuzzy-search library.

The very first step of integrating any external library is to look at the code (often examples) to grasp an idea of what is to be achieved from R. Below is an example from the official documentation. First, an array of two `books` is defined; this is later used to test the search. Then another array of options is defined. This should at the very least include the key(s) that should be searched; here it is set to search through the title and authors. Then, the fuse object is initialised based on the array of books and the options. Finally, the `search` method is used to retrieve all books, the title or author of which partially match the term `tion`.

```
// books to search through
var books = [{
  'ISBN': 'A',
  'title': "Old Man's War",
  'author': 'John Scalzi'
}, {
  'ISBN': 'B',
  'title': 'The Lock Artist',
  'author': 'Steve Hamilton'
}]
```

---

[4]https://fusejs.io/

```
const options = {
  // Search in `author` and in `title` array
  keys: ['author', 'title']
}

// initialise
const fuse = new Fuse(books, options)

// search 'tion' in authors and titles
const result = fuse.search('tion')
```

With some understanding of what is to be reproduced in R, we can import the library with the `source` method, which takes a `file` argument that will accept a path or URL to a JavaScript file to source. Below we use the handy CDN (Content Delivery Network) to avoid downloading a file.

```
uri <- paste0(
  "https://cdnjs.cloudflare.com/ajax/",
  "libs/fuse.js/3.4.6/fuse.min.js"
)
engine$source(uri)
#> [1] "true"
```

You can think of it as using the `script` tag in HTML to source (`src`) said file from disk or CDN.

```
<html>
  <head>
    <script
      src='https://cdnjs.cloudflare.com/.../fuse.min.js'>
    </script>
  </head>
  <body>
  </body>
</html>
```

Now onto replicating the array (list) which we want to search through, the `books` object used in a previous example. As already observed, this is in essence, how V8 stores data frames in the environment. Below we define a data frame of books that looks similar and load it into the engine.

```r
books <- data.frame(
  title = c(
    "Rights of Man",
    "Black Swan",
    "Common Sense",
    "Sense and Sensibility"
  ),
  id = c("a", "b", "c", "d")
)


engine$assign("books", books)
```

Then again, we can make sure that the data frame was turned into a row-wise JSON object.

```r
cat(engine$eval("JSON.stringify(books, null, 2);"))
#> [
#>   {
#>     "title": "Rights of Man",
#>     "id": "a"
#>   },
#>   {
#>     "title": "Black Swan",
#>     "id": "b"
#>   },
#>   {
#>     "title": "Common Sense",
#>     "id": "c"
#>   },
#>   {
#>     "title": "Sense and Sensibility",
#>     "id": "d"
#>   }
#> ]
```

Now we can define options for the search; we don't get into the details of fuse.js here as this is not the purpose of this book. You can read more about the options in the examples section[5] of the site. We can mimic the format of the JSON options shown on the website with a simple list and assign that to a new variable in the engine. Note that we wrap the title in a list to ensure

---

[5]https://fusejs.io/#Examples

it is converted to an array of length 1: list("title") should be converted to a
["title"] array and not a "title" scalar.

```javascript
// JavaScript
var options = {
  keys: ['title'],
  id: 'id'
}
```

```r
# R
options <- list(
  keys = list("title"),
  id = "id"
)


engine$assign("options", options)
```

Then we can finish the second step of the online examples, instantiate a fuse.js
object with the books and options objects, then do a search, the result of
which is assigned to an object which is retrieved in R with get.

```r
engine$eval("var fuse = new Fuse(books, options)")
engine$eval("var results = fuse.search('sense')")
engine$get("results")
#> [1] "d" "c"
```

A search for "sense" returns a vector of ids where the term "sense" was found;
c and d or the books *Common Sense, Sense and Sensibility*. We could perhaps
make that last code simpler using the call method.

```r
engine$call("fuse.search", "sense")
#> [1] "d" "c"
```

18.4   NPM Packages

We can also use npm[6] packages, though not all will work. NPM is Node's package manager, or in a sense Node's equivalent of CRAN.

To use NPM packages we need browserify[7], a node library to bundle all dependencies of an NPM package into a single file, which can subsequently be imported in V8. Browserify is itself an NPM package, and therefore requires Node.js to be installed. The reason browserify is required will be covered in more depth in chapter 20, in essence, NPM assumes disk access to load dependencies in require() (JavaScript) statements. This will not work with V8. Browserify will bundle all the files that comprise an NPM module into a single file that does not require disk access.

You can install browserify globally with the following the g flag. Once Node.js installed, browserify can be installed from the *terminal* (not R console) with the npm command.

```
npm install -g browserify
```

We can now "browserify" an npm package. To demonstrate, we will use ms[8], which converts various time formats to milliseconds. First, we install the npm package.

```
npm install ms
```

Then we browserify it. From the terminal, the first line creates a file called in.js which contains global.ms = require('ms'); we then call browserify on that file specifying ms.js as output file. The require function in JavaScript is used to import files, require('ms') imports ms.js, it's to some extend like source("ms.R").

```
echo "global.ms = require('ms');" > in.js
browserify in.js -o ms.js
```

We can now source ms.js with V8. Before we do so we ought to look at example

---

[6]https://www.npmjs.com/

[7]http://browserify.org/

[8]https://github.com/zeit/ms

code to see what has to be reproduced using V8. Luckily the library is very straightforward: it includes a single function for all conversions, e.g.: `ms('2 days')` to convert two days in milliseconds.

```r
library(V8)

ms <- v8()
ms$source("ms.js")
```

Then using the library simply consists of using `eval` or preferably `call` (for cleaner code and data interpretation to R).

```r
ms$eval("ms('2 days')")
#> [1] "172800000"
ms$call("ms", "2s") # 2 seconds
#> [1] 2000
```

18.5 Use in Packages

In this section, we detail how one should go about using V8 in an R package. If you are not familiar with package development you can skip ahead. We start by creating a package called "ms" that will hold functionalities we explored in the previous section on NPM packages.

```r
usethis::create_package('ms')
```

The package is going to rely on V8 so it needs to be added under `Imports` in the `DESCRIPTION` file, then again this can be done with usethis as shown below.

```r
# add V8 to DESCRIPTION
usethis::use_package("V8")
```

The package should also include the external library `ms.js` browserified from the NPM package, which should be placed it in the `inst` directory. Create it and place the `ms.js` file within the latter.

```
dir.create("inst")
```

As explored, the core of the V8 package is the execution environment(s) that are spawned using the v8 function. One could perhaps provide a function that returns the object created by v8, but it would not be convenient: this function would need to be called explicitly by the users of the package, and the output of it would need to be passed to every subsequent function. Thankfully there is a better way.

Instead, we can use the function .onLoad, to create the execution environment and import the dependency when the package is loaded by the user.

You can read more about this function in Hadley Wickham's *Advanced R* book[9]. This is, in effect, very similar to how the Python integration of R, reticulate[10] (Ushey et al., 2020), is used in packages[11]. This function is often placed in a zzz.R file.

```
# zzz.R
ms <- NULL

.onLoad <- function(libname, pkgname){
  ms <<- V8::v8()
}
```

At this stage the package's directory structure should look similar to the tree below.

```
.
├── DESCRIPTION
├── NAMESPACE
├── R
│   └── zzz.R
└── inst
    └── ms.js
```

Now the dependency can be sourced in the .onLoad function. We can locate the files in the inst directory with the system.file function.

---

[9]http://r-pkgs.had.co.nz/r.html
[10]https://rstudio.github.io/reticulate
[11]https://rstudio.github.io/reticulate/articles/package.html

```
# zzz.R
ms <- NULL


.onLoad <- function(libname, pkgname){
  ms <<- V8::v8()

  # locate dependency file
  dep <- system.file("ms.js", package = "ms")
  ms$source(dep)
}
```

We can then create a `to_ms` function. It will have access to the `ms` object we instantiated in `.onLoad`.

```
#' @export
to_ms <- function(string){
  ms$call("ms", string)
}
```

After running `devtools::document()` and installing the package with `devtools::install()`, it's ready to be used.

```
ms::to_ms("10 hrs")
#> [1] 36000000
```

19

Machine Learning

In this chapter, we build a package that, via V8, wraps ml.js[1], a library that brings machine learning to JavaScript. It covers quite a few models; we only include one: the linear regression. This is an interesting example because it reveals some proceedings that one is likely to run into when creating packages using V8.

```
const x = [0.5, 1, 1.5, 2, 2.5];
const y = [0, 1, 2, 3, 4];

const regression = new ml.SimpleLinearRegression(x, y);
```

19.1 Dependency

We start by creating a package and add the V8 package as dependency.

```
usethis::create_package("ml")
usethis::use_package("V8")
```

Then we create the inst directory in which we place the ml.js file downloaded from the CDN.

```
dir.create("inst")
download.file(
  "https://www.lactame.com/lib/ml/4.0.0/ml.min.js",
  "inst/ml.min.js"
)
```

---

[1]https://github.com/mljs/ml

With the dependency downloaded, one can start working on the R code. First, a new V8 context needs to be created and the ml.js file needs to be imported into it.

```
# zzz.R
ml <- NULL

.onLoad <- function(libname, pkgname){
  ml <<- V8::v8()
  mljs <- system.file("ml.min.js", package = "ml")
  ml$source(mljs)
}
```

19.2   Simple Regression

The "simple linear regression"[2] consists of a simple function that takes two arrays. We can thus create a function that takes two vectors, x, and y, and runs the regression.

```
#' @export
ml_simple_lm <- function(y, x){
  # assign x and y
  ml$assign("x", x)
  ml$assign("y", y)

  # run regression
  ml$eval(
    "var regression = new ML.SimpleLinearRegression(x, y);"
  )

  # return results
  ml$get("regression")
}
```

Then we can document and load the model to the function.

---

[2]https://github.com/mljs/regression-simple-linear

```
ml_simple_lm(cars$speed, cars$dist)
## $name
## [1] "simpleLinearRegression"
##
## $slope
## [1] 0.1655676
##
## $intercept
## [1] 8.283906
```

This works but has a few issues, namely running two or more regression internally will override the variable `regression` in the V8 context. Let us demonstrate by implementing a function to predict.

```
#' @export
ml_predict <- function(x){
  ml$call("regression.predict", x)
}
```

We then document and load the functions to run two regressions in a row then observe the issue. Unlike R, the model we created only exists in JavaScript, unlike the `lm`, the function `ml_simple_lm` does not return the model. Therefore, `ml_simple_lm` does not distinguish between models, unlike `predict`, which takes the model as the first argument and runs the prediction on it.

This implementation of ml.js is indeed very dangerous.

```
# first model
ml_simple_lm(cars$speed, cars$dist)
ml_predict(15)
## 25.18405

# overrides model
ml_simple_lm(1:10, runif(10))

# produces different predictions
ml_predict(15)
## 10.76742
```

The package ml currently under construction should emulate R in that respect; the `ml_simple_lm` should return the model, which should be usable with the `predict` function. In order to do so, we are going to need to track regressions internally in V8 so the `ml_simple_lm` returns a proxy of the model that `predict` can use to predict on the intended model.

In order to track and store regressions internally, we are going to declare an empty array when the package is loaded.

```
# zzz.R
ml <- NULL

.onLoad <- function(libname, pkgname){
  ml <<- V8::v8()
  mljs <- system.file("ml.min.js", package = "ml")
  ml$source(mljs)
  ml$eval("var regressions = [];")
}
```

Then, one can track regressions by creating an R object, which is incremented every time `ml_simple_lm` runs; this can be used as a variable name in the JavaScript `regressions` array declare in `.onLoad`. This variable name must be stored in the object we intend to return so the `predict` method we will create later on can access the model and run predictions. Finally, in order to declare a new method on the `predict` function, we need to return an object bearing a unique class. Below we use `mlSimpleRegression`.

```
counter <- new.env(parent = emptyenv())
counter$regressions <- 0

#' @export
ml_simple_lm <- function(y, x){
  counter$regressions <- counter$regressions + 1

  # assign variables
  ml$assign("x", x)
  ml$assign("y", y)

  # address
  address <- paste0("regressions['", counter$regressions, "']")

  # create regression
  code <- paste0(
    address, " = new ML.SimpleLinearRegression(x, y);"
  )
```

```
  ml$eval(code)

  # retrieve and append address
  regression <- ml$get(address)
  regression$address <- address

  # create object of new class
  structure(
    regression,
    class = c("mlSimpleRegression", class(regression))
  )
}
```

Then one can implement the `predict` method for `mlSimpleRegression`. The function uses the `address` of the model to run the JavaScript `predict` method on that object.

```
#' @export
predict.mlSimpleRegression <- function(object, newdata, ...){
  code <- paste0(object$address, ".predict")
  ml$call(code, newdata)
}
```

We can then build and load the package to it in action.

```
library(ml)

# first model
model_cars <- ml_simple_lm(cars$speed, cars$dist)

# second model
model_random <- ml_simple_lm(1:10, runif(10))

predict(model_random, 15)
#> [1] -130.8514
predict(model_cars, 15)
#> [1] 10.76742
```

19.3 Exercises

There are too many great JavaScript libraries that would be great additions to the R ecosystem but perhaps try and integrate one of these below. They are simple yet exciting and thus provide ideal first forays into what this part of the book explained.

- chance.js[3] - random generator
- currency.js[4] - helpers to work with currencies

---

[3]https://github.com/chancejs/chancejs
[4]https://github.com/scurker/currency.js

Part V

Robust JavaScript

20

Managing JavaScript

Thus far, all of the JavaScript code written in this book was placed directly in the file that was imported in the front end, be it htmlwidgets or Shiny-related code. While this works for the smaller projects, it is bound to lead to headaches for the larger ones.

It's the same problem one faces when writing R code. While a small script of 300 lines of code will do the job, a large script of 10,000 lines quickly becomes unmanageable. Therefore when tackling more extensive projects, the R programmer will turn to solutions that enforce a specific file structure and provide utilities to harmonise how those files work together. Some of these solutions may include the drake (Landau, 2021a) or targets (Landau, 2021b) packages, both of which provide tools to manage complex workflows. Another method often used is to build the project as an R package, thereby enforcing a particular structure and enabling reproducibility, unit tests, and more.

The issues mentioned above are also a concern in JavaScript, though here one has to consider additional pitfalls. Like R, JavaScript is a continually evolving language, but while R code written on version 4.0.0 will likely run fine on version 3.0.0, it is not precisely the case for JavaScript. As the language evolves and changes, web browsers have to keep up to support any new feature brought by new releases.

Therefore JavaScript code that is written on the latest version may not run on all browsers. Also, consider that even if the latest versions of Google Chrome and Mozilla Firefox tend to support the latest JavaScript, users who visit your Shiny applications or use your htmlwidgets may not have their browsers up to date.

In JavaScript, code mismanagement might be exacerbated because it often relies on other files such as CSS, JSON, images, and more, making it challenging to build robust projects. Moving an image from one folder to another or removing a CSS file may break an entire JavaScript project.

Also, in JavaScript, code size matters: the smaller the file, the faster it will load in the browser. JavaScript files are reduced in size with a process called "minification," which consists of removing all unnecessary characters, such as spaces, from a JavaScript file to obtain a "minified" version that fits on a single line, which is smaller in size. This is because humans cannot read or write minified code; try removing all line breaks and spaces from your R scripts if you think otherwise, then imagine JavaScript minification takes it a step further.

Finally, since R is rather strict, packages enforce a specific structure; JavaScript does not come with such restrictions off the shelf. Therefore, it's even more tempting for the developer to take shortcuts and make a mess of their projects.

Combine all of the above and software that involves JavaScript can quickly become poorly structured and cumbersome. Moreover, considering all these potential issues as one writes code is unsustainable as it dramatically increases the cognitive load and ultimately distracts from writing code itself: it's just too much to consider. Thankfully some tools have been invented over the years to help JavaScript developers manage all of these matters. These tools differ slightly from one another, they each have their pros and cons, but all have the same goal: making JavaScript projects more robust and manageable.

Grunt[1] describes itself as a "the JavaScript task runner," and will carry minification, compilation, unit testing, linting, and more. There is also Parcel[2] a web application bundler that will minify and bundle (and more) JavaScript code. However, the one we shall use in this part of the book is webpack[3], as it is very similar to Grunt and is one of the most popular.

20.1   Example

There are admittedly few R packages that make use of such technology, though it must be said that many could greatly benefit from it. Given its size and complexity a package such as Shiny, however, could probably not do without it.

Shiny makes use of Grunt, the source code that comprises all of the JavaScript required to run the front end (inputs, outputs, modals, etc.) is in the `srcjs` directory which can be found on the official GitHub repository.[4] This folder includes a multitude of JavaScript files the names of which indicate the code they encompass; `input_binding_checkbox.js`, `modal.js`, etc.

These files are processed by Grunt which, using the `Gruntfile.js` configuration file in the `tools` directory, creates multiple bundles that it places in the `inst` folder of the package.

---

[1] https://gruntjs.com/

[2] https://parceljs.org/

[3] https://webpack.js.org/

[4] https://github.com/rstudio/shiny

20.2 Transpiling

As new functionalities are made available in JavaScript, with every modern version web browsers have to keep pace and support running said functionalities. First, this is not always the case, major web browsers such as Google Chrome, Mozilla Firefox, and Safari generally do a decent job of keeping up, but one can never count on the individuals using those to do keep their browsers up to date.

Imagine building a large htmlwidgets for a client only to discover that for IT security reasons all their company laptops run a particular version of a web browser that does not support critical functionalities the widget relies upon.

Ensuring that the JavaScript code can run on most browsers is no trivial task. The best way to do so *is not* to write outdated JavaScript code that all browsers should support, the solution is actually to use a Babel.[5] This transpiler will convert "ECMAScript 2015+ code into a backwards-compatible version of JavaScript." This way one can use the latest JavaScript, even before browsers officially support it, and transpile it with Babel to obtain a JavaScript file that will run on any browser that supports ECMAScript 2015 (JavaScript version released in 2015).

20.3 Minification

Web browsers always need to load the files necessary to render a webpage, be it a static website, a Shiny application, or a standalone widget. Loading those files can take critical time and make the loading of a web application slow. Therefore it is good practice to reduce the size of those files. This includes compressing images, so they are smaller in size and load faster but also "minifying" CSS and JavaScript code.

When writing code, us humans like to use comprehensible variable names, line breaks, spaces, and other things that help make things clear and readable. Machines, however, do not need any of that; as long as the code is valid, it will run.

---

[5]https://babeljs.io/

```
// global variable
let number = 41;

// my hello function
function hello(my_variable){
    let total = number + my_variable;
    console.log(total);
}
```

Minification is the process of removing all of the "syntactic sugar" that is unnecessary to obtain JavaScript code that fits in a single line and makes for a smaller file. See the example given here where the code above is minified to get the code below, note that the comment was removed and even some variable names have changed to be shorter.

```
let number=41;function hello(e){let l=number+e;console.log(l)}
```

The minified files of a library tend to end in `.min.js` though minified code can very well be placed in a `.js` file.

20.4   Bundling and Modules

Managing the structure of JavaScript projects can be tricky. One does not want to place 2,000 lines of code in a single file, but splitting JavaScript code into multiple files is complicated.

While writing an R package, one is free to organise the functions in different files as their content (functions, data, etc.) is ultimately loaded into the global with `library` by the user of the package.

In JavaScript, one does not have the luxury of writing code across different files to then call `library()` in the web browser, so all the functions, and variables are available. In this paradigm, individual files have to be loaded separately in the browser (as shown below).

```
<script src="utils.js"></script>
<script src="main.js"></script>
```

While this may be fine for two or three files, it quickly gets out of hand as one has to remember to import those in the correct order. In the above example, variables declared in `main.js` cannot be used in `utils.js`, unless we change the order of the import in which case something else will likely break.

It's therefore essential to use tools that allow splitting JavaScript programs into modules (to write programs in different files), manage the dependencies between these files, then "bundle" those correctly into one or more files destined to be imported in the browser.

20.5 Decoupling

One thing that might become apparent from the previous sections is the idea of decoupling the final code that makes it into the web browser from the code we write. It is thanks to this decoupling that we can write easy-to-read JavaScript code on the latest version across multiple files to then run the various processes of transpiling, bundling, and minifying to obtain code that is more efficient and robust for the browser.

This may appear like a lot to manage. Thankfully we can use the aforementioned webpack software to take care of all these procedures for us. There is nonetheless a gentle learning curve to make use of it as it involves multiple new concepts.

Moreover, webpack does not limit itself to the previously-mentioned processes. It will also take care of other things such as removing "dead code," functions or variables that are declared but not used, and it allows integrating CSS and other files in JavaScript itself, and so much more.

20.6 NPM

Another new piece of software that we need to be introduce is Node's Package Manager, hereafter referred to as NPM. As indicated by the name, it's a package manager for Node.js, or translated for the R user it's Node's loose equivalent of CRAN. One first significant difference is that while CRAN performs very rigorous checks on any package submitted, NPM does not; one can publish almost anything.

Notice how every dependency used in this book had to be either found through a CDN or manually downloaded, only to be imported in the final document. Again, this is useful for the smaller projects but may become a hindrance when multiple dependencies have to be managed and updated, added and removed, etc.

NPM has wholly changed how dependencies can be managed and imported in JavaScript. It is designed for Node.js code, but many (if not all) libraries that are meant to run in web browsers are published on NPM: it's just too convenient.

NPM, combined with the decoupling, and bundling covered in previous sections, enables managing dependencies much more sensibly, and efficiently. So one can, for instance, import only certain functions from an external library rather than the whole, thereby further reducing the size of the final bundle of JavaScript files.

20.7   With R

To be clear, it's not always necessary to involve webpack and NPM into a project; these can take some time to set up and be excessive for a smaller project. It's good to be familiar with them as one might want to make use of those in larger projects.

Webpack and NPM were not designed with R in mind, so there are some potential issues to consider when using it in packages and Shiny applications.

In the following chapter, we discover how to include both webpack and NPM to make more robust Shiny applications, widgets, and other packages that involve JavaScript.

21

Discover Webpack and NPM

In this chapter, we discover how to feature webpack and NPM in a straightforward Shiny project. The idea is not to build a complex application, only to find out how one might go about bringing them into an R project and observe some of their benefits (and potential issues).

There is a lot of depth to NPM and webpack; we only touch upon the surface here so we can obtain a basic setup for a Shiny application. We'll eventually go into slightly more detail as this part of the book progresses, but it will by no means fully explore the realm of webpack. It's always a good idea to take a look at the official documentation[1] to get a better picture of the technology.

21.1 Installation

As Node's Package Manager, a working installation of Node.js, is required, NPM ships with it. A bit like R in a sense where the package manager also comes with the language, install R, and you can install packages from CRAN with `install.packages`. The same applies here, install Node and you can install NPM packages from the command line.

We are only going to use Node.js *indirectly*, some of its functionalities and its package manager. This is not about building Node applications.

Below are some directions on how to install Node.js. In the event this does not work or you encounter issues please refer to the official website[2].

21.1.1 Mac OS

On Mac OS, the easiest way is via homebrew.

---

[1]https://webpack.js.org/
[2]https://nodejs.org/en/

DOI: 10.1201/9781003134046-21

```
brew update
brew install node
```

Otherwise there is also an installer[3] available.

21.1.2 Ubuntu

With Ubuntu one can install it straight from the package manager.

```
sudo apt install nodejs
```

21.1.3 Windows

Download and install the official executable[4] or use chocolatey[5].

```
cinst nodejs.install
```

Or use scoop[6].

```
scoop install nodejs
```

21.1.4 Other

If you are on another OS or Linux distro check the official, very concise guide[7] to install from various package managers.

21.2 Set Up the App

Let us first put together a simple Shiny application that will serve as a basis for including webpack and npm. Create a new directory and in it place a file called app.R containing a very simple application.

---

[3]https://nodejs.org/en/download/
[4]https://nodejs.org/en/download/
[5]https://chocolatey.org/
[6]https://scoop.sh/
[7]https://nodejs.org/en/download/package-manager/

```
library(shiny)

ui <- fluidPage(
  h1("A shiny app")
)

server <- function(...) {}

shinyApp(ui, server)
```

21.3 Initialise NPM

With a simple application, one can initialise NPM. This could be translated into the equivalent of starting a new project in R. This is done from the command line, *from the root of the directory* that you want to use as a project.

Whereas in R we previously used the usethis package to create packages with create_package or projects with create_project, NPM does not create the initial empty directory where the project will be created; you have to create the directory first then initialise a new project.

An NPM project can be initialised with the command npm init, which when run prompts the user with a few questions, such as the name of the project, the license to use, etc. These have little importance for what we do here but will matter if you decide to publish the package on NPM. One can also pass the "yes" flag to the function to skip those questions: npm init -y.

This creates a package.json file, which is loosely equivalent to the DESCRIPTION of an R package; it includes information on the dependencies of the project, the version of the project, and more.

We will revisit this file later in the chapter. At this stage ensure you have run npm init (with or without the -y flag) from the root of the project (where the app.R file is located).

21.4 Installing NPM Packages

Unless the R programmer uses packages such as renv (Ushey, 2020) or packrat
(Ushey et al., 2018) then packages are installed globally on the machine,
running `install.packages("dplyr")` installs a single version of dplyr across the
entire device. Because CRAN is strict and its packages subsequently stable, it
tends not to be too much of an issue. Packages submitted to are checked for
reverse dependencies (other packages that depend on it) to see if the submission
could cause problems downstream.

However, NPM does no such thing with packages that are submitted. Therefore
the developer has to be more careful about dependencies, particularly versioning
as packages can dramatically change from one version to the next. Thus
it makes sense that NPM out-of-the-box advocates and provides tools to
encapsulate projects. It is *not recommended,* to install NPM packages globally.
NPM projects (the directory where `npm init` was run) come bundled with the
equivalent of renv/packrat.

Installing Node packages also takes place at the command line with the `install`
command followed by the name of the package to install, e.g.: `npm install`
`nameOfPackage.`

As mentioned, it is rarely a good idea to install packages globally at the
exception of very few packages, such as command-line applications used across
the machine. As an example, the docsify-cli[8] package for documentation
generation can safely be installed globally as it is used at the command line in
projects that don't necessarily use NPM. This can be achieved with the `-g` flag
that stands for "global": `npm install docsify-cli -g.`

There are two other scopes on which packages can be installed. NPM allows
distinguishing between packages that are needed to develop the project and
packages that are needed in the final product being built.

R does not come with such convenience but it could perhaps be useful. For
instance throughout the book we used the usethis package to develop packages
from setting it up to adding packages to the DESCRIPTION file, and more. Perhaps
one would like to make this a "developer" dependency so that other developers
that pull the package from GitHub have usethis installed and readily available.
The advantage is that this dependency would not be included in the final
product, that is, usethis is not required to use the package (only to develop it)
and therefore is not installed by the user.

---

[8]`https://docsify.js.org/`

As stated in the previous chapter, file size matters in JavaScript; it is, therefore, crucial that dependencies that are used only for development are not included in the final JavaScript file. With NPM this can be done by using the `--save-dev` flag, e.g.: `npm install webpack --save-dev` to install webpack. This is how it will be eventually installed as it is needed to prepare the final product (minify, bundle, etc.) but is not required to run the bundled file(s).

Finally, there are the "true" dependencies, those that are needed in the output we're creating. For instance, were we to rebuild the gio widget using NPM we could install it with `npm install giojs --save` because this dependency will be required in the output file we produce.

Before moving on to the next section, let us install webpack and its command-line interface as developer dependencies.

```
npm install webpack webpack-cli --save-dev
```

Notice that this updated the `package.json` file and created the `package-lock.json` file as well as a `node_modules` directory to obtain the following structure.

```
.
├── app.R
├── node_modules
├── package-lock.json
└── package.json
```

The directory `node_modules` actually holds all the dependencies, and it will grow in size as you add more, it's important that this directory is not pushed to whatever version control system you happen use (GitHub, Bitbucket, Gitlab).

Exclude the `node_modules` directory from your version control (Git or otherwise)

The dependencies are anyway not needed as one can pull the project without the `node_modules` then from the root of the project run `npm install` to install the dependencies that are listed in the `package.json` file. We can indeed observe that this file was updated to include `webpack` and `webpack-cli` as `devDependencies`, at the bottom of the file.

```
{
  "name": "name-of-your-project",
  "version": "1.0.0",
```

```
"description": "",
"main": "index.js",
"scripts": {
  "test": "echo \"Error: no test specified\" && exit 1"
},
"keywords": [],
"author": "",
"license": "ISC",
"devDependencies": {
  "webpack": "^5.2.0",
  "webpack-cli": "^4.1.0"
}
}
```

The `package-lock.json` file is automatically generated and *should not be edited manually*. It describes the exact tree of all the dependencies. If you installed a package by mistake, you could uninstall it with `npm uninstall nameOfPage`.

Recap

- Install packages globally with `npm install package -g`
- Install developer dependencies with `npm install package --save-dev`
- Install dependencies required in the output with `npm install package --save`
- Uninstall packages with `npm uninstall package`

21.5   Entry Point and Output

In general, an NPM project with webpack will make use of an `src` directory where the source code is placed and a `dist` directory (for distributed) where the bundled source code will be placed; we'll see how to change these defaults later on. It will eventually be necessary as the `src` directory in R packages is reserved for compiled code (e.g., C++) and therefore cannot be used to place JavaScript files.

It will not be a problem here as we are not building a package.

```
dir.create("src")
```

Webpack will then require at least one "entry point." An entry point is an input file in the src directory that webpack will use as a source to produce the bundle. Let's create the go-to "hello world" of JavaScript; the snippet below creates the index.js file with a basic vanilla JavaScript alert.

```
writeLines("alert('hello webpack!')", "src/index.js")
```

The next section on configuration will detail precisely how to indicate to webpack that this is indeed the entry point it should use.

21.6 Configuration File

Webpack comes with a configuration file, webpack.config.js. Though for a larger project it is advised to split it into multiple configuration files (more on that later). This file can include numerous options, plugins, and other settings to customise how webpack transforms the entry point into an output, only some of which will be explored in this book as there are too many to cover.

Below is probably the most straightforward configuration file one may create. At the bare minimum, the configuration file will need to have an entry point specified; in this case, the previously-created index.js file. If no output path is specified, then webpack will produce it at dist/main.js automatically.

```
// webpack.config.js
module.exports = {
  entry: './src/index.js'
};
```

The module.exports line may confuse; it is covered in a later section on *importing and exporting* variables and functions.

21.7 NPM scripts

NPM scripts allow automating development tasks such as running unit tests, serving files, and more, we'll set it up to run webpack. The scripts are placed in the package.json file and are terminal commands.

By default `npm-init` creates the following `test` script, which echoes (prints) a message stating that no unit tests were set up.

```
"scripts": {
  "test": "echo \"Error: no test specified\" && exit 1"
}
```

This script can be run from the terminal by typing `npm run test`. Those commands always follow the same pattern: `npm run` followed by the name of the script, in this case, `test`.

Adding the script to run webpack is very straightforward, we can add an entry called `build` that runs `webpack`.

```
"scripts": {
  "test": "echo \"Error: no test specified\" && exit 1",
  "build": "webpack"
}
```

So running `npm run build` produces the output file from the entry point file. However, we will modify this slightly in the next section as more features of webpack are uncovered.

21.8   Source maps

We will improve upon the previous section so we can run webpack on two different modes: one for production and one for development.

Since the output of webpack is any number of files bundled into one, it can make debugging more difficult. When files `a.js`, `b.js`, and `c.js` are bundled into `dist/main.js`, the stack trace will point to errors in `dist/main.js`, which is not helpful as the developer needs to know in which original file the bug lies.

Therefore, webpack comes with a "development" mode that allows including the "source map," which maps the compiled code to the source files. This way, when an error or warning is raised JavaScript is able to point to the original line of code that causes it.

There are again many different ways to set this up in the configuration file as the source map can be placed in the bundled file itself, in another file, and more. However, the easiest way is probably to specify the mode using webpack's CLI tool. The source maps are optional as these make the output larger and one wants to keep this output as small as possible for it to load as fast as possible in web browsers. Those will thus only be used while developing the project to trace back errors and warnings but will not be included in the final output for production.

Below we modify the scripts placed in the package.json file so two different scripts can be run: one for development and another for production.

```
"scripts": {
  "test": "echo \"Error: no test specified\" && exit 1",
  "build-prod": "webpack --mode=production",
  "build-dev": "webpack --mode=development"
}
```

This allows running npm run build-prod to produce the production bundle and npm run build-dev to produce the development version that includes the source map.

21.9   Bundle

One can then bundle the code using the scripts that we defined to produce the output bundle. Since we have not specified any output in webpack's configuration file, it will create it at the default location dist/main.js.

```
npm run build-prod
```

We can then include the output of webpack in the Shiny application to test that all works well.

```
library(shiny)

mainJs <- htmltools::htmlDependency(
  name = "main",
  version = "1.0.0",
  src = "./dist",
```

```
  script = c(file = "main.js")
)

ui <- fluidPage(
  mainJs,
  h1("A shiny app")
)

server <- function(...) {}

shinyApp(ui, server)
```

Running the above launches the app, which presents the `alert()` that was placed in the `index.js` source file.

This makes for a great start but is not precisely interesting; in the following sections, we elaborate on this basic configuration to make better use of webpack's feature and produce something much more fun.

21.10   Internal Dependencies

Let's install a dependency and use it in our Shiny application.

We'll install mousetrap,[9] a library to handle key-strokes. We're going to use it to hide the UI of the Shiny application behind a secret pass-phrase; it will only be revealed after it has been typed. This can be done by observing a specific set of key-strokes with mousetrap and set a Shiny input value when that particular sequence is typed.

This is by no means a safe way to secure an application!

Though it is certainly not a real-world example, it is educational and quite a bit of fun.

The first thing to do is to install the mousetrap dependency; as indicated on the GitHub README[10] it can be obtained from NPM.

---

[9] https://github.com/ccampbell/mousetrap
[10] https://github.com/ccampbell/mousetrap

```
npm install mousetrap --save
```

Note that we use `--save` as mousetrap will need to be included in the output we create, it's not a library we import for development purposes.

21.11 External Dependencies

If dependencies with webpack have to be installed from NPM it begs the question; what about dependencies that are already included in the project and are not available on NPM.

For instance, this project is intended to work with a Shiny application, which comes bundled with , and the Shiny JavaScript library. First, the Shiny javaScript library is not available on NPM. Second installing it would result in duplicating dependencies, which is hardly best practice. Thankfully webpack comes with a simple mechanism to handle these cases; external dependencies can be added to the configuration file under `externals`.

```
module.exports = {
  entry: './src/index.js',
  externals: {
    shiny: 'Shiny'
  }
};
```

The above will allow importing the `shiny` object in scripts, which is needed to set the input value with `Shiny.setInputValue`; hence `shiny` must be accessible in webpack. Let us delve into the import/export mechanism.

21.12 Import and Export

To demonstrate how webpack enables modularising code, we will not place all the code in the `index.js` file. We create two other files: `secret.js` and `input.js`. The first will contain the pass-phrase and the second will have the code to handle the key strokes via mousetrap and set the Shiny input. This will enable using the pass-phrase in multiple places without duplicating code.

```
file.create("src/input.js")
file.create("src/secret.js")
```

Therefore, as shown in Figure 21.1, the entry point `index.js` needs to import the `input.js` file, which itself imports the pass-phrase from `secret.js`.

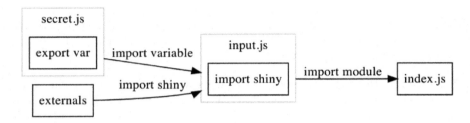

FIGURE 21.1: Webpack with Shiny

Again, there are multiple ways to import and export modules, functions, variables, etc. This book will use the ES6 syntax as recommended by webpack.[11] Though this mechanism is present in other languages, such as Python (where it somewhat resembles ES6), it will take some getting used to for R programmers as though this language features some form of import (`library()`) and export (`@export` roxygen2 tag), this differs significantly from how it works in webpack. This is, however, key to using webpack, as it is what ultimately enables the creation of modules that make code more robust.

There are two different kinds of exports and imports possible, "named" and "default." We shall cover them in that order.

21.12.1 Named

Let's place the variable `secret` in the `secret.js` file. As a reminder, this variable will have to be imported by in another file (`input.js`) where it will be used to check if the pass-phrase typed by the user is correct.

Declaring the variable itself does not change, we use the keyword `let` to declare a variable named `secret` that holds the pass-phrase. The issue is that with

---

[11]https://webpack.js.org/api/module-methods/#es6-recommended

webpack, this variable will be internal to the file where it is declared. However, we ultimately want to import that variable in another file. To do so, we can place the keyword `export` in front of the declaration to indicate that this variable is exported from the file. Note that this will also work with functions and classes, and other objects.

Placing `export` in front of an object constitutes a *named export*; the `secret.js` file explicitly exports the variable named `secret`.

```
export let secret = 's e c r e t';
```

Then this variable can be imported in the `input.js` file. The named export in `secret.js` comes with a corresponding named import in `input.js` to import the variable named `secret`; this is indicated by the curly braces. Note that again we include the path to the file (`./secret.js`), importing from `secret.js` without the path will fail.

```
import { secret } from './secret.js';
```

The curly braces are used for named imports as multiple such variables or functions can then be imported, e.g., `import { foo, bar } from './file.js';` to import the named exports `foo` and `bar` from `file.js`.

21.12.2 Default

An alternative would be to use a default export. A file can have a default export; said default could be a variable, a function, a list, or any number of things but *there can only be a single default export per file*.

```
// declare
let secret = 's e c r e t';

// export
export default secret;
```

Rather interestingly, because multiple variables can be declared on a single line (e.g., `var a,b,c;`) but only a single default can exist, the default export and vriable declaration cannot be placed on a single line.

```
// invalid
export default secret = 's e c r e t';

// valid
var x = 0,
    y = true;

export default {x, y}
```

This only applies to variables as only a single function can be declared by line so declaring a function and its default export on athe same line is valid.

```
// valid
export default function sayHello() {
  alert("Hello!")
};
```

Importing default exports in other files resembles all too much the syntax of named imports, which may lead to confusion: it's essentially the same omitting the curly braces.

```
import secret from './secret.js';
```

21.12.3   Wrap-up

We'll be using a named export method in secret.js. The same general logic can be applied to import the external dependency Shiny as well as mousetrap.

```
import Shiny from 'shiny';
import { secret } from './secret.js';
import Mousetrap from 'mousetrap';

Mousetrap.bind(secret, function() {
  Shiny.setInputValue('secret', true);
});
```

Finally, remember to import input.js in the entry point index.js.

```
// index.js
import './input.js';
```

This can then be bundled with `npm run bundle-prod`, which will start at the entry point (`index.js`) observe that it imports the file `input.js`, then look at that file and see that it imports `secret.js`; webpack builds this dependency tree and includes all that is needed in the bundle.

This can be used in the Shiny application, which we modify so it listens to the `secret` input and only when that input is set renders a plot and a message.

```r
library(shiny)

mainJs <- htmltools::htmlDependency(
  name = "main",
  version = "1.0.0",
  src = "./dist",
  script = c(file = "main.js")
)

ui <- fluidPage(
  mainJs,
  p("Type the secret phrase"),
  uiOutput("hello"),
  plotOutput("plot")
)

server <- function(input, output) {
  output$hello <- renderUI({
    req(input$secret)
    h2("You got the secret right!")
  })

  output$plot <- renderPlot({
    req(input$secret)
    hist(cars$speed)
  })
}

shinyApp(ui, server)
```

Once the application is launched the user can type the phrase `secret` to see the content of the application (see Figure 21.2).

Type the secret phrase

You got the secret right!

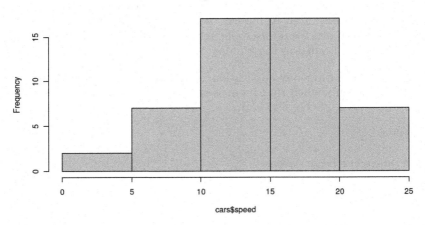

FIGURE 21.2: Mousetrap example

That is it for this chapter. As stated multiple times there is far more depth to webpack, but this is outside the scope of this book, instead in the next chapter we discover an easier way to set up such projects and make R and webpack work in a more seamless fashion.

22

Webpack with R

In the previous chapter, we put together a simple Shiny application using NPM and webpack. Hopefully, it hinted at some of the powerful things webpack can do but also revealed a downside: the overhead in merely creating the project. Moreover, the configuration will change depending on what the project is (application, package, etc.).

In this chapter, we discover the packer[1] (Coene, 2020) R package, which provides many convenience functions to create and manage R projects that make use of webpack and NPM.

```r
install.packages("packer")
```

22.1 Principles of packer

There are a few principles that the packer package follows strictly.

1. It only aspires to become a specialised usethis for working with JavaScript and R. As such, it takes inspiration from other packages such as htmlwidgets and devtools.
2. It will never become a dependency to what you create. It's in a sense very much like an NPM "developer" dependency; it's used to develop the project but does not bring any additional overhead to what you're building.
3. It should not interfere with the mission of webpack to build more robust JavaScript code. Therefore, packer only builds on top of already, strict R structures, namely packages (where golem can be used to create Shiny applications).

---

[1] https://github.com/JohnCoene/packer

DOI: 10.1201/9781003134046-22

22.2 Scaffolds

Packer is comprised of surprisingly few functions; the most important ones are in the `scaffold` family. The term scaffold was borrowed from the htmlwidgets package, which features the function `scaffoldWidget` (already used in this book). The idea of scaffolds in packer is very similar to the `scaffoldWidget` function: they set up the basic structure for projects.

Whilst htmlwidgets only allows creating scaffolds for widgets; packer allows creating scaffolds for several things, namely:

- Widgets with `scaffold_widget`
- Shiny inputs with `scaffold_input`
- Shiny outputs with `scaffold_output`
- Shiny extensions with `scaffold_extension`
- Golem applications with `scaffold_golem`

This gives a few powerful functions that correctly set up webpack. These will build the necessary file structure and configuration depending on the scaffold and the context (whether it is a basic package, a golem application, a package with an existing scaffold, etc.)

One can use multiple scaffolds in a single package or Shiny application.

Packer goes beyond merely setting up webpack and NPM; it will also create the necessary R functions, roxygen documentation, and examples, so every scaffold is fully functional out-of-the-box.

With some variations that will be explored in the coming sections, packer's `scaffold` functions generally do the following:

- Initialise npm with `npm init` and prefills the `package.json` file
- Install webpack and its CLI with `npm install webpack webpack-cli --save-dev`
- Creates three webpack configuration files: `webpack.common.js`, `webpack.prod.js`, and `webpack.dev.js`
- Creates the `srcjs` directory for the JavaScript source code
- Creates raw JavaScript files within the `srcjs` directory, e.g.: `index.js`
- Creates the R functions (if necessary)
- Adds the necessary NPM scripts to `package.json`
- Adds all relevant files to the `.Rbuildignore` and `.gitignore` files

- Adds relevant dependencies to the DESCRIPTION, e.g.: shiny when scaffolding an input
- Finally, it (optionally) opens interesting files to develop the project in the IDE

In the following sections, we unpack some of this as we explore a specific scaffold.

22.3 Inputs

In a previous chapter, we explored how to build custom Shiny inputs. Here, we'll use the packer package to produce a Shiny button that increments at every click; hence we create a package called "increment."

```
usethis::create_package("increment")
```

From the root of the package, we scaffold a custom input. Notably, this takes a name argument, which is used as names for the various files, functions, and modules it creates so choose it with care. The function prints some information about the operations it executes.

When run from an interactive session, packer also opens the most pertinent files in the default editor or IDE.

```
packer::scaffold_input("increment")
```

```
── Scaffolding shiny input ──────────────────── increment ──
Initialiased npm
Created srcjs/inputs directory
Created inst/packer directory
webpack, webpack-cli, webpack-merge installed with scope dev
Created srcjs/config directory
Created webpack config files
Created 'input' module
Created srcjs/index.js
Created R file and function
Added npm scripts
```

```
— Adding files to '.gitignore' and '.Rbuildignore' —

Setting active project to '/javascript-for-r/code/increment'
Adding '^srcjs$' to '.Rbuildignore'
Adding '^node_modules$' to '.Rbuildignore'
Adding '^package\\.json$' to '.Rbuildignore'
Adding '^package-lock\\.json$' to '.Rbuildignore'
Adding '^webpack\\.dev\\.js$' to '.Rbuildignore'
Adding '^webpack\\.prod\\.js$' to '.Rbuildignore'
Adding '^webpack\\.common\\.js$' to '.Rbuildignore'
Adding 'node_modules' to '.gitignore'

— Adding packages to Imports —

Adding 'shiny' to Imports field in DESCRIPTION
• Refer to functions with `shiny::fun()`
Adding 'htmltools' to Imports field in DESCRIPTION
• Refer to functions with `htmltools::fun()`

— Scaffold built —

Run `bundle` to build the JavaScript files
```

The scaffold creates the file structure below. Notice that increment was used as the name of some files and that packer creates three webpack configuration files; one for development, another for production, and a third that contains configuration shared across those two modes.

It created one R file, increment.R, which contains the exported input function named incrementInput. It also created the inst/packer directory, which is currently empty but will eventually contain the bundled JavaScript file(s).

The function also initialised NPM, which created the node_modules directory, as well as the package.json and package-lock.json, packer also added the necessary scripts to package.json so one should not need to interact with those files directly.

Finally, it also created the srcjs directory containing core JavaScript files to produce the input binding.

```
.
├── DESCRIPTION
├── NAMESPACE
├── R
```

```
|    ├── increment.R
├── inst
|    └── packer
├── node_modules
|    └── ...
├── package.json
├── srcjs
|    ├── config
|    ├── inputs
|    └── index.js
├── webpack.common.js
├── webpack.dev.js
└── webpack.prod.js
```

In the following sections, we break down those files to better understand what packer scaffolded and how to use it.

22.4   R file

The R file contains the `incrementInput` function. Notably, the function contains the necessary dependency, although it currently looks for a file that is yet created (we'll bundle the JavaScript later). Also, of importance is the class attribute set for the input: `incrementBinding`. As you might remember, this class will be referenced in the JavaScript binding's `find` method.

```r
incrementInput <- function(inputId, value = 0){

  stopifnot(!missing(inputId))
  stopifnot(is.numeric(value))

  dep <- htmltools::htmlDependency(
    name = "incrementBinding",
    version = "1.0.0",
    src = c(file = system.file("packer", package = "increment")),
    script = "increment.js"
  )

  tagList(
    dep,
    tags$button(
```

```
      id = inputId,
      class = "incrementBinding btn btn-default",
      type = "button",
      value
    )
  )
}
```

Note that packer does not use the namespace of functions (e.g., `shiny::tagList`). Instead, it uses the roxygen2 tags to import the necessary functions: `@importFrom Shiny tags tagList`. Rather nicely, packer also created an example in the roxygen documentation. We'll run this later after we've bundled the JavaScript.

```
#' @examples
#' library(shiny)
#'
#' ui <- fluidPage(
#'   incrementInput("theId", 0)
#' )
#'
#' server <- function(input, output){
#'
#'   observeEvent(input$theId, {
#'     print(input$theId)
#'   })
#'
#' }
#'
#' if(interactive())
#'   shinyApp(ui, server)
```

22.5   JavaScript Files

In the `srcjs/inputs` directory, packer created `increment.js`. This code contains the JavaScript binding for the increment button. As a reminder, one is not limited to a single scaffold. We could scaffold another input, the JavaScript binding of which would be placed alongside this file, also in `srcjs/inputs`.

```javascript
import $ from 'jquery';
import 'shiny';

$(document).on("click", "button.incrementBinding",
  function(evt) {
    // evt.target is the button that was clicked
    var el = $(evt.target);

    // Set the button's text to its current value plus 1
    el.text(parseInt(el.text()) + 1);

    // Raise an event to signal that the value changed
    el.trigger("change");
  }
);

var incrementBinding = new Shiny.InputBinding();

$.extend(incrementBinding, {
  find: function(scope) {
    return $(scope).find(".incrementBinding");
  },
  getValue: function(el) {
    return parseInt($(el).text());
  },
  setValue: function(el, value) {
    $(el).text(value);
  },
  subscribe: function(el, callback) {
    $(el).on("change.incrementBinding", function(e) {
      callback();
    });
  },
  unsubscribe: function(el) {
    $(el).off(".incrementBinding");
  }
});

Shiny.inputBindings.register(
  incrementBinding, "increment.incrementBinding"
);
```

The `srcjs/index.js` file was also created; it imports the JavaScript binding detailed above with `import './inputs/increment.js';`. Notably, by default, packer does not bundle all of these files into one; `index.js` is only populated for convenience in the event one would want to change this behaviour. Instead, packer uses `srcjs/inputs/increment.js` as an entry point. It will handle multiple entry points, so every input, output, widgets, etc. are bundled separately. This is done so one can import those dynamically.

22.6   Bundle

You can then run `packer::bundle` to bundle the JavaScript. The entry points and output directories will depend on the scaffold, Shiny inputs' bundles are placed in the `inst/packer` directory unless this was run from a golem application, in which case the output is automatically changed to golem's standard.

```
packer::bundle()
```

By default packer will bundle the files for production, this can be managed with the functions `packer::bundle_dev()` and `packer::bundle_prod()`.

Once the JavaScript is bundled, we can install or load the package with `devtools::load_all` and use the example that was created for us to test the input.

```
library(shiny)

ui <- fluidPage(
  incrementInput("theId", 0)
)

server <- function(input, output){

  observeEvent(input$theId, {
    print(input$theId)
  })

}

if(interactive())
  shinyApp(ui, server)
```

No code was written, yet we have a fully-functional input! We'll leave ir at this: it's not only meant to create increment buttons, but this sets up a solid base for the developer to customise the code and conveniently create a different input.

It is worth noting that we built a Shiny input from within a package, this is meant to be exported and used in Shiny applications elsewhere, but were one to run these same steps from a golem application packer would adapt the output path so that the input can be used directly in the application.

23

Webpack Advanced

We're about to cover slightly more advanced uses of NPM and webpack with R using packer. These involve using an NPM dependency to develop a widget and use Vue.js[1] and Bootstrap 4 to power the front end of a Shiny application.

Those will make for more concrete cases to bring webpack into your workflow, and also enable explaining more advanced topics only thus far briefly touched upon, such as transpiling.

23.1 Widgets

The widget scaffold, like all other scaffolds, must be run from within the root of a package. To demonstrate we'll write a widget for the countup[2] library that allows animating numbers.

First, we create a package which we name `counter`. You can name the package differently but avoid naming it `countup`. We will later have to install the external dependency also named `countup` and NPM does not allow a project named X to use a dependency also named X.

```
usethis::create_package("counter")
```

From the root of the package, we scaffold the widget with `scaffold_widget`, which prints out some information on what packer exactly does.

```
packer::scaffold_widget("countup")
```

— Scaffolding widget ——————————————————————— countup —

---

[1]https://vuejs.org/
[2]https://github.com/inorganik/countUp.js/

DOI: 10.1201/9781003134046-23

```
Bare widget setup
Created srcjs directory
Initialiased npm
webpack, webpack-cli, webpack-merge installed with scope dev
Created srcjs/config directory
Created webpack config files
Created srcjs/modules directory
Created srcjs/widgets directory
Created srcjs/index.js
Moved bare widget to srcjs
Added npm scripts


── Adding files to '.gitignore' and '.Rbuildignore' ──


Setting active project to '/Projects/countup'
Adding '^srcjs$' to '.Rbuildignore'
Adding '^node_modules$' to '.Rbuildignore'
Adding '^package\\.json$' to '.Rbuildignore'
Adding '^package-lock\\.json$' to '.Rbuildignore'
Adding '^webpack\\.dev\\.js$' to '.Rbuildignore'
Adding '^webpack\\.prod\\.js$' to '.Rbuildignore'
Adding '^webpack\\.common\\.js$' to '.Rbuildignore'
Adding 'node_modules' to '.gitignore'


── Adding packages to Imports ──


Adding 'htmlwidgets' to Imports field in DESCRIPTION
● Refer to functions with `htmlwidgets::fun()`


── Scaffold built ──


Run `bundle` to build the JavaScript files
```

Importantly, it runs `htmlwidgets::scaffoldWidget` internally, there is thus no need to run this function. About the widget itself, there is very little difference between what `htmlwidgets::scaffoldWidget` and `packer::scaffold_widget`. While, if you remember, the initial scaffold of htmlwidgets includes a simple function to display a message in HTML using `innerText`. The scaffold produced by packer differs only in that this message is displayed in `<h1>` HTML tags. That is so it can, from the get-go, demonstrate how to modularise a widget. We'll cover that in just a minute, before we do so, bundle the JavaScript and run the `counter` function to observe the output it generates.

```
packer::bundle()
devtools::load_all()
countup("Hello widgets!")
```

This indeed displays the message in `<h1>` HTML tags, now onto unpacking the structure generated by packer. We'll skip the R code to keep this concise as nothing differs from a standard widget on that side. Instead, we'll focus on the JavaScript code in the `srcjs` directory. First, in the `srcjs/widgets` directory, one will find the file `countup.js`. This file contains the code that produces the widget.

At the top of the file are the imports. First, it imports `widgets`, which is the htmlwidgets *external dependency,* second it imports the function `asHeader` from the `header.js` file in the `modules` directory.

```
import 'widgets';
import { asHeader } from '../modules/header.js';

HTMLWidgets.widget({

  name: 'countup',

  type: 'output',

  factory: function(el, width, height) {

    // TODO: define shared variables for this instance

    return {

      renderValue: function(x) {

        // TODO: code to render the widget, e.g.
        el.innerHTML = asHeader(x);

      },

      resize: function(width, height) {

        // TODO: code to re-render the widget with a new size

      }
```

```
    };
  }
});
```

The `header.js` file includes the `asHeader` function, which accepts an `x` argument that is used to create the `<h1>` message. This function is exported.

```
const asHeader = (x) => {
  return '<h1>' + x.message + '</h1>';
}


export { asHeader };
```

We will make changes to the JavaScript so that instead of displaying the message as text, it uses the aforementioned countup library to animate a number. The first order of business is to install countup. Here we use packer to do so; the function call below is identical to running `npm install countup --save` from the terminal.

```
packer::npm_install("countup", scope = "prod")
```

We will not need the `header.js` file. We can delete it and in its stead create another file called `count.js`. This file will include a function that uses countup to animate the numbers. It should accept 1) the id of the element where countup should be used, and 2) the value that countup should animate. This function called `counter` is, at the end of the file, exported.

```
import { CountUp } from 'countup.js';

function counter(id, value){
  var countUp = new CountUp(id, value);
  countUp.start();
}

export { counter };
```

We need to add the import statement to bring in the `counter` function and run it in the `renderValue` method. Packer also added the htmlwidgets external dependency, which is imported below with `import 'widgets'`.

Because we left the *R function* `countup` untouched, we have to use the default
`message` variable it accepts. Ideally, this argument in the R function should be
renamed to something more adequate.

```
import 'widgets';
import { counter } from '../modules/count.js';

HTMLWidgets.widget({

  name: 'countup',

  type: 'output',

  factory: function(el, width, height) {

    // TODO: define shared variables for this instance

    return {

      renderValue: function(x) {

        counter(el.id, x.message);

      },

      resize: function(width, height) {

        // TODO: code to re-render the widget with a new size

      }

    };
  }
});
```

Finally, the JavaScript bundle can be generated with `packer::bundle()`, install
the package or run `devtools::load_all()`, and test that the widget works!

```
countup(12345)
```

That hopefully is a compelling example to use NPM and webpack to build
widgets. It could even be argued that it is easier to set up; dependencies are

much more manageable; nothing has to be manually downloaded; it will be easier to update them in the future, etc.

23.2 Shiny with Vue and Bootstrap 4

In this example, we create a Shiny application that uses Vue.js[3] and Bootstrap 4 in the front end. As you may know, Shiny ships with Bootstrap version 3, not 4 (the latest at the time of writing this).

23.2.1 Setup

If you prefer using React[4], know that it is also supported by webpack and packer. Vue is a framework to create user interfaces that, like React, make much of the front end work much more straightforward. It reduces the amount of code one has to write, simplifies business logic, enables reactivity, and much more.

Since packer only allows placing scaffolds in R packages, the way one can build Shiny applications is using the golem package. Golem is an opinionated framework to build applications *as R packages*. Writing Shiny applications as R packages brings many of the advantages that packages have to Shiny applications: ease of installation, unit testing, dependency management, etc.

```r
install.packages("golem")
```

After installing golem from CRAN, we can create an application with the `golem::create_golem` function; it's very similar to `usethis::create_package`, only it prepares a package specifically to build Shiny applications.

```r
golem::create_golem("vuer")
```

From within a golem application, one uses a scaffold specifically designed for this with `scaffold_golem`. Note that this does not mean other scaffolds will not work, custom Shiny inputs and outputs can also be created with `scaffold_input` and `scaffold_output`, respectively. The `scaffold_golem` function takes two core

---

[3]https://vuejs.org/
[4]https://reactjs.org/

arguments; vue and react. Setting either of these to TRUE will prepare a scaffold specifically designed to support either Vue or React.

The reason these arguments exist is that webpack requires further configuration that can be tricky to set up manually. Moreover, Vue supports (but does not require) .vue files; these can hold HTML, JavaScript, and CSS. One can think of such files as similar to Shiny modules; they encapsulate a part of the logic of the application for easy modularisation.

When the vue argument is set to TRUE in scaffold_golem, the function does follow the usual procedure or initialising NPM, creating the various files, and directories, but in addition configures two loaders and the vue plugin.

Loaders are transformers, they scan files in the srcjs directory and pre-process them. That allows using, for instance, the Babel compiler that will transform the latest version of JavaScript into code that every browser can run. This compiler is very often used, including here, to compile Vue code. Since Vue allows placing CSS in .vue files, another loader is required; one that will look for CSS and bundle it within the JavaScript file.

Plugins are a feature of webpack that allow extending its functionalities; there is one for Vue, which the function will install and configure for you.

Also, when creating a scaffold for vue or react, one can choose whether to rely on the CDN, in which case they are installed as developer dependencies, or install them for production, in which case they are included in the bundle. It defaults to using the CDN; this is often advised as the CDN will serve the required files faster.

The scaffold also sets up webpack with Babel, the transpiler that allows writing the latest JavaScript, and ensures it will run on (almost) any web browser. Hence, we can use ES6 notation in places.

```
packer::scaffold_golem(vue = TRUE)
```

```
— Scaffolding golem ─────────────────────────────
Initialiased npm
webpack, webpack-cli, webpack-merge installed with scope dev
Added npm scripts
Created srcjs directory
Created srcjs/config directory
Created webpack config files

— Adding files to '.gitignore' and '.Rbuildignore' —

Setting active project to '/Projects/vuer'
Adding '^srcjs$' to '.Rbuildignore'
```

```
Adding '^node_modules$' to '.Rbuildignore'
Adding '^package\\.json$' to '.Rbuildignore'
Adding '^package-lock\\.json$' to '.Rbuildignore'
Adding '^webpack\\.dev\\.js$' to '.Rbuildignore'
Adding '^webpack\\.prod\\.js$' to '.Rbuildignore'
Adding '^webpack\\.common\\.js$' to '.Rbuildignore'
Adding 'node_modules' to '.gitignore'


── Vue loader, plugin & dependency ──


babel-loader installed with scope dev
Added loader rule for 'babel-loader'
@babel/core, @babel/preset-env installed with scope dev
vue installed with scope dev
vue-loader, vue-template-compiler installed with scope dev
Added loader rule for 'vue-loader' and 'vue-template-compiler'
style-loader, css-loader installed with scope dev
Added loader rule for 'style-loader' and 'css-loader'
Created R/vue_cdn.R containing `vueCDN()` function
Added alias to srcjs/config/misc.json


── Babel config file ──


Created '.babelrc'
Adding '^\\.babelrc$' to '.Rbuildignore'


── Template files ──


Added srcjs/Home.vue template
! Place the following in your shiny ui:
tagList(
  vueCDN(),
  div(id = "app"),
  tags$script(src = "www/index.js")
)


── Scaffold built ──


Run `bundle` to build the JavaScript files
```

Note the first instruction that was printed in the console; it states a tagList must be placed in the Shiny UI of the application. It imports the Vue dependency via the CDN with vueCDN(), which is a function created by packer, creates a <DIV> with an id attribute of app that will be used as root of the Vue application; where the application generated by Vue will be placed. It also imports the bundled JavaScript (`index.js`). So let us place that in the Shiny UI, which is in the `R/app_ui.R` file.

```r
app_ui <- function(request) {
  tagList(
    golem_add_external_resources(),
    fluidPage(
      tagList(
        vueCDN(),
        div(id = "app"),
        tags$script(src = "www/index.js")
      )
    )
  )
}
```

To ensure all is correct up to this point we can test the application; the JavaScript can be bundled with `packer::bundle()` then the app tested by running `run_app()`.

```r
packer::bundle()
devtools::load_all()
run_app()
```

23.2.2 Bootstrap 4 Installation

Next, we can install Bootstrap 4; we'll use bootstrap-vue[5] which contains a lot of Bootstrap 4 components for Vue. We won't be using any CDN here, so we install those dependencies as production.

```r
packer::npm_install("bootstrap-vue", "bootstrap", scope = "prod")
```

This will cause some issues though, as the Shiny application will have two different versions of Bootstrap, the default version 3 and version 4 from the bundle. We need to remove Bootstrap 3.

---

[5] https://bootstrap-vue.org/

```
app_ui <- function(request) {
  tagList(
    golem_add_external_resources(),
    # remove default bootstrap 3
    htmltools::suppressDependencies("bootstrap"),
    fluidPage(
      tagList(
        vueCDN(),
        div(id = "app"),
        tags$script(src = "www/index.js")
      )
    )
  )
}
```

23.2.3   Vue Code

Let us now explore the contents of srcjs and code a basic functionality. It's relatively straightforward; it consists of two files. The first, index.js, creates the Vue application and places it in the div(id = "app"). The code for the app itself is in a .vue file, which it imports with import App from "./Home.vue";.

```
import Vue from "vue";
import App from "./Home.vue";

new Vue({
  el: "#app",
  template: "<App/>",
  components: { App }
});
```

The first order of business is to import the Bootstrap dependencies that were installed and "use" them in the application. We don't explain this in great detail here as much of it is specific to Vue and is thus outside the scope of this book.

```
import Vue from "vue";
import { BootstrapVue, IconsPlugin } from 'bootstrap-vue'
// import dependencies
```

```
import 'bootstrap/dist/css/bootstrap.css'
import 'bootstrap-vue/dist/bootstrap-vue.css'
import App from "./Home.vue";

// attach dependencies
Vue.use(BootstrapVue)
Vue.use(IconsPlugin)

new Vue({
  el: "#app",
  template: "<App/>",
  components: { App }
});
```

The Home.vue file is where the meat of the application is placed. By default,
packer creates an app that just displays a message.

```
<template>
  <p>{{ greeting }} powered by Vue!</p>
</template>

<script>
module.exports = {
  data: function() {
    return {
      greeting: "Shiny"
    };
  }
};
</script>

<style scoped>
p {
  font-size: 2em;
  text-align: center;
}
</style>
```

Below we make changes to the application, so it features a Boostrap 4 text
input. After having entered some text and hitting enter the text entered is
displayed below and cleared from the input. We also provide a button that
sends the input data to the R server.

```
<template>
  <div>
    <b-form-input
      v-model="inputText"
      placeholder="Enter your name"
      @keyup.enter="processText">
    </b-form-input>
    <b-button
      @click="processText"
      variant="outline-primary">
      Button
    </b-button>
    <h2>Your name is {{ text }}</h2>
  </div>
</template>

<script>
module.exports = {
  data: function() {
    return {
      text: '',
      inputText: ''
    };
  },
  methods: {
    processText: function(){
      this.text = this.inputText // set text var
      Shiny.setInputValue('text', this.text);
      this.inputText = '' // remove input once entered
    }
  }
};
</script>
```

Finally, we can bundle the JavaScript, and run the application to obtain Figure 23.1.

```
packer::bundle()
devtools::load_all()
run_app()
```

Hello Vue!

Button

Your name is Hello Vue!

FIGURE 23.1: Shiny application with Vue and Bootstrap 4

Note how little code was written in order to provide these functionalities. It is one of the most powerful features of frameworks like Vue and React. They are not necessary; this could have been coded in vanilla JavaScript, but would admittedly require much more (difficult to read) code.

Part VI

Closing Remarks

24

Conclusion

The book covered a few topics and introduced many concepts probably new to many; these are sometimes difficult to fully understand at first and will likely require some more exploring. So here, we wrap up with some heartening thoughts to encourage the reader to persevere, as much of what was covered in the book requires practice to be fully comprehended.

24.1 Performances

As mentioned in the opening of this book some of the code presented is not entirely optimal as we need to make use of R as much as possible to limit any confusion caused by the fresh-to-the-eyes JavaScript code. In places, this involves making use of R where it is in fact not needed, namely within Shiny applications where data is sent from the front end to R, only to be sent back to JavaScript again. For instance, when the click of a button triggers something in the Shiny UI but uses the server as an intermediary, oftentimes it is not necessary to involve the server.

This, though, has little impact on performances in most cases, and can be improved upon by not involving R in the process and handling everything in the front end with JavaScript. This was judged outside the scope of the book as it focuses on interactions between the two languages. Nonetheless, somewhat interestingly, the book covered all the code necessary to do so, only not in the same section or chapter. It might, therefore, take some practice to make the connection.

```javascript
// toggle an input at the click of a button
$('#button').on('click', function(){
  $('#input').toggle();
});
```

Note, however, that placing much of the business logic server-side rather than on the front end might create more secure applications since said logic remains internal and cannot be interfered with.

24.2   Trial and Error

We hope the book demonstrated how well JavaScript works with R, as well as how much of a difference it has the potential to have on your data science projects. Making JavaScript work *for R* is fascinating because they are so far removed from one another; while one excels at data wrangling and statistics, the other runs in the browser and focuses on aesthetics and functionalities of web pages.

However, being so different, JavaScript introduces numerous concepts likely to be new to many R developers. The only way one can truly grasp how it all works, and become at ease with using custom JavaScript code in Shiny applications and packages, is to practice. Repeated trial and error is fundamental to approaching new programming languages and notions.

Some small exercises were scattered at the end of significant parts of the book; you are encouraged to attempt some of them. Like interactive visualisations? Try to build one for a straightforward library!

24.3   Functionality and UX

A lot of JavaScript in the browser is about designing user experiences and better functionalities for users of your data product–never overlook those.

We don't say of a chart with pleasing aesthetics "that's just a pretty plot"; a great chart does a better job of communicating insights, and the same is true of many other data products, including web applications. Hopefully, the learnings contained in this book will help you create much more engaging and compelling products thanks to JavaScript.

Sadly, in the fields where R is popular, things like aesthetics, or great user experience are perceived as superfluous. You might be told that "the only thing that matters is the analysis or the model"; anything else is often seen as make-up to cover up flawed science. This could not be further from the truth. First, put to rest the false dichotomy that it's either a great front end or a great back end; both can (and must) be done. Second, while it could be said that spending two hours looking for a fun colour palette for a chart is "a waste of time," spending the same amount of time developing new JavaScript functionalities to allow users to interrogate your model better, or visualise (and therefore understand) the outcome of an analysis is not. Also, if your visualisations and web applications are engaging, users will gladly spend more time clicking away, interrogating results, and using your product.

Bibliography

Allaire, J., Xie, Y., McPherson, J., Luraschi, J., Ushey, K., Atkins, A., Wickham, H., Cheng, J., Chang, W., and Iannone, R. (2021). *rmarkdown: Dynamic Documents for R*. R package version 2.7.

Bache, S. M. and Wickham, H. (2020). *magrittr: A Forward-Pipe Operator for R*. R package version 2.0.1.

Chang, W. and Borges Ribeiro, B. (2018). *shinydashboard: Create Dashboards with 'Shiny'*. R package version 0.7.1.

Chang, W., Cheng, J., Allaire, J., Sievert, C., Schloerke, B., Xie, Y., Allen, J., McPherson, J., Dipert, A., and Borges, B. (2021a). *shiny: Web Application Framework for R*. R package version 1.5.0.9005.

Chang, W., Cheng, J., Dipert, A., and Borges, B. (2021b). *websocket: 'WebSocket' Client Library*. R package version 1.3.2.

Cheng, J. (2016). *crosstalk: Inter-Widget Interactivity for HTML Widgets*. R package version 1.0.0.

Cheng, J. and Chang, W. (2021). *httpuv: HTTP and WebSocket Server Library*. R package version 1.5.5.

Cheng, J., Sievert, C., Chang, W., Xie, Y., and Allen, J. (2021). *htmltools: Tools for HTML*. R package version 0.5.1.9000.

Coene, J. (2020). *packer: An Opinionated Framework for Using 'JavaScript'*. https://github.com/JohnCoene/packer, https://packer.john-coene.com.

Coene, J. (2021a). *echarts4r: Create Interactive Graphs with 'Echarts JavaScript' Version 5*. https://echarts4r.john-coene.com/, https://github.com/JohnCoene/echarts4r.

Coene, J. (2021b). *waiter: Loading Screen for 'Shiny'*. https://waiter.john-coene.com/, https://github.com/JohnCoene/waiter.

Cooley, D. (2020). *jsonify: Convert Between 'R' Objects and Javascript Object Notation (JSON)*. R package version 1.2.1.

Guyader, V., Fay, C., Rochette, S., and Girard, C. (2020). *golem: A Framework for Robust Shiny Applications*. R package version 0.2.1.

Henry, L. and Wickham, H. (2020). *purrr: Functional Programming Tools*. R package version 0.3.4.

Inc, F., Weststrate, M., Russell, K., and Dipert, A. (2020). *reactR: React Helpers*. R package version 0.4.3.

Kunst, J. (2020). *highcharter: A Wrapper for the 'Highcharts' Library*. R package version 0.8.2.

Landau, W. M. (2021a). *drake: A Pipeline Toolkit for Reproducible Computation at Scale*. R package version 7.13.1.

Landau, W. M. (2021b). *targets: Dynamic Function-Oriented 'Make'-Like Declarative Workflows*. R package version 0.1.0.

Lin, G. (2020). *reactable: Interactive Data Tables Based on 'React Table'*. R package version 0.2.3.

Ooms, J. (2020). *jsonlite: A Simple and Robust JSON Parser and Generator for R*. R package version 1.7.2.

Sievert, C., Parmer, C., Hocking, T., Chamberlain, S., Ram, K., Corvellec, M., and Despouy, P. (2021). *plotly: Create Interactive Web Graphics via 'plotly.js'*. R package version 4.9.3.

Strayer, N., Luraschi, J., and Allaire, J. (2020). *r2d3: Interface to 'D3' Visualizations*. R package version 0.2.5.

Teucher, A. and Russell, K. (2020). *rmapshaper: Client for 'mapshaper' for 'Geospatial' Operations*. R package version 0.4.4.

Ushey, K. (2020). *renv: Project Environments*. R package version 0.9.3.

Ushey, K., Allaire, J., and Tang, Y. (2020). *reticulate: Interface to 'Python'*. R package version 1.18.

Ushey, K., McPherson, J., Cheng, J., Atkins, A., and Allaire, J. (2018). *packrat: A Dependency Management System for Projects and their R Package Dependencies*. R package version 0.5.0.

Vaidyanathan, R. (2013). *rCharts: Interactive Charts using Javascript Visualization Libraries*. R package version 0.4.5.

Vaidyanathan, R., Xie, Y., Allaire, J., Cheng, J., Sievert, C., and Russell, K. (2020). *htmlwidgets: HTML Widgets for R*. R package version 1.5.3.

Wickham, H. (2019). *stringr: Simple, Consistent Wrappers for Common String Operations*. R package version 1.4.0.

Wickham, H. (2020). *testthat: Unit Testing for R.* R package version 2.3.2.

Wickham, H. and Bryan, J. (2021). *usethis: Automate Package and Project Setup.* R package version 2.0.1.

Wickham, H., Chang, W., Henry, L., Pedersen, T. L., Takahashi, K., Wilke, C., Woo, K., Yutani, H., and Dunnington, D. (2020a). *ggplot2: Create Elegant Data Visualisations Using the Grammar of Graphics.* R package version 3.3.2.

Wickham, H., Danenberg, P., Csárdi, G., and Eugster, M. (2020b). *roxygen2: In-Line Documentation for R.* R package version 7.1.1.

Wickham, H., Hester, J., and Chang, W. (2020c). *devtools: Tools to Make Developing R Packages Easier.* R package version 2.3.2.

Xie, Y., Cheng, J., and Tan, X. (2020). *DT: A Wrapper of the JavaScript Library 'DataTables'.* R package version 0.16.

You, E. and Russell, K. (2020). *vueR: 'Vuejs' Helpers and 'Htmlwidget'.* R package version 0.5.2.

Index

asset, 30, 31, 158, 159, 189, 192, 209

CDN, 34, 68, 105, 170, 226, 259, 280, 313

CRAN, 3, 7, 12, 18, 41, 136, 255, 262, 279, 281, 284, 312

dependency, 12, 14, 15, 29, 32, 34, 36, 37, 49, 60, 66, 68–70, 78, 103–105, 107, 108, 122, 157, 159, 169, 170, 188, 189, 192, 193, 201–205, 209, 213, 226, 264, 267, 268, 279, 280, 283–286, 290, 291, 297, 301, 307, 310, 316

DOM, 24, 27, 110, 156, 191

ECMA, 25, 277

environment, 18, 25, 26, 30, 91, 107, 174, 256, 259, 264, 278

export, 16, 17, 32, 91, 107, 175, 225, 287, 292, 293, 300, 305, 310

HTML, 23, 31–33, 54, 58, 59, 63–65, 68, 75, 93, 193, 223, 237, 259, 308

htmlwidgets, 41, 49, 54, 57, 62, 64, 75, 76

jQuery, 68–70, 73, 184, 185, 194, 215, 216, 291

JSON, 19–22, 50, 60, 62–64, 82, 100, 109, 110, 133, 148, 204, 217, 241, 260

pipe, 43, 91, 246

R markdown, 34, 35, 54, 60, 96, 103, 108, 110, 118

React, 8

resize, 62, 96, 97

RStudio, 11–13, 24, 29, 58, 59, 96, 118, 189

scaffold, 57, 62, 68, 188, 298, 300, 307, 308, 313

scope, 24, 25

serialise, 19, 20, 54, 63, 73, 84–88, 99–101, 109, 150, 195, 217, 240

Shiny, 11, 29, 30, 32, 41, 47, 54, 58, 101, 103, 113, 139, 141, 151, 155, 156, 159, 166, 177, 180, 209

visualisation, 41, 51, 52, 54, 63, 89, 94, 104, 122, 136, 232, 239, 246

Vue, xxi, 8, 307, 312–316, 319

web browser, 7, 11, 23, 27, 58, 59, 63, 96, 97, 100, 110, 111, 139, 155, 225, 234, 277, 280, 289

WebSocket, 139, 141, 147, 157, 223